T0180215

Springer Aerospace Technology

Series Editors

Sergio De Rosa, DII, University of Naples Federico II, NAPOLI, Italy

Yao Zheng, School of Aeronautics and Astronautics, Zhejiang University, Hangzhou, Zhejiang, China

The series explores the technology and the science related to the aircraft and spacecraft including concept, design, assembly, control and maintenance. The topics cover aircraft, missiles, space vehicles, aircraft engines and propulsion units. The volumes of the series present the fundamentals, the applications and the advances in all the fields related to aerospace engineering, including:

- structural analysis,
- aerodynamics,
- aeroelasticity,
- aeroacoustics,
- flight mechanics and dynamics
- orbital maneuvers,
- avionics,
- systems design,
- materials technology,
- launch technology,
- payload and satellite technology,
- space industry, medicine and biology.

The series' scope includes monographs, professional books, advanced textbooks, as well as selected contributions from specialized conferences and workshops.

The volumes of the series are single-blind peer-reviewed.

To submit a proposal or request further information, please contact:

Mr. Pierpaolo Riva at pierpaolo.riva@springer.com (Europe and Americas)
Mr. Mengchu Huang at mengchu.huang@springer.com (China)

The series is indexed in Scopus and Compendex

More information about this series at http://www.springer.com/series/8613

Dmitry Alexandrovich Zatuchny ·
Ruslan Nikolaevich Akinshin ·
Nina Ivanovna Romancheva ·
Igor Viktorovich Avtin ·
Yury Grigorievich Shatrakov

Noise Resistance Enhancement in Aircraft Navigation and Connected Systems

 Springer

Dmitry Alexandrovich Zatuchny
Moscow, Russia

Nina Ivanovna Romancheva
Moscow, Russia

Yury Grigorievich Shatrakov
Saint Petersburg, Russia

Ruslan Nikolaevich Akinshin
Moscow, Russia

Igor Viktorovich Avtin
Moscow, Russia

ISSN 1869-1730   ISSN 1869-1749 (electronic)
Springer Aerospace Technology
ISBN 978-981-16-0632-8   ISBN 978-981-16-0630-4   (eBook)
https://doi.org/10.1007/978-981-16-0630-4

This Springer imprint is published by the registered company Springer Nature Singapore Pte Ltd.
The registered company address is: 152 Beach Road, #21-01/04 Gateway East, Singapore 189721, Singapore

Introduction

The present time is characterized by the almost ubiquitous widespread introduction of modern satellite technologies designed to accurately and reliably locate all types of vehicles. The solution to this problem is of particular relevance in relation to civil aircraft, when it is necessary to ensure the proper reliability and quality of navigation signals received from satellite systems. The efficiency and accuracy of determining the coordinates can be significantly influenced by both objective circumstances related to weather conditions, industrial activities and the nature of the underlying surface of the Earth and subjective factors resulting from interstate and regional military conflicts and the actions of militarized terrorist groups. In this case, there is a real threat of a shutdown of navigation systems and even the radio-technical suppression of signals.

Classically, the noise stability of an aircraft's navigation system is the ability to provide stable reception of useful signals even if it is affected by natural and man-made interference. In this case, the main thing under these conditions is the possibility of high-quality functioning of the system in the presence of interference at the input of receiving channels. Noise immunity refers to the ability of a system to maintain tactical characteristics at a given level when conducting electronic warfare, to function successfully (effectively) in the presence of natural interference and jamming, and also to counteract reconnaissance of its radio signals by the enemy.

Thus, the concept of noise immunity includes noise stability and secrecy. In accordance with the requirements of the International Civil Aviation Organization (ICAO), all radio frequency parameters shall be open to all interested airspace users. Therefore, the terms of noise stability and noise immunity can be considered identical for civil aviation.

In general, the noise stability of the consumer receiving device is determined by the logic and operating stages of the receiving equipment, the structure and algorithms for processing navigation radio signals, their energy characteristics, the number of channels in the consumer receiver, methods of integrating navigation information, and the time–frequency structure of the interference at the receiver input.

It should be noted that the main factor in achieving superiority over competitors in the twenty-first century, according to some experts, will be the availability of cutting-edge information technologies as the basis for the development and application of highly efficient global monitoring, communication, and control systems. Therefore, the relevance of scientific research to solve the problems of ensuring information superiority has increased dramatically.

All of this is directly related to the problem of the smooth and accurate functioning of modern flight-navigation systems and connected systems of civil aircraft, which currently have to perform not only their direct tasks but are, in fact, information and control and information and computer systems, as well as information security systems of civil aircraft.

To improve the noise stability of civil aircraft under modern conditions, it is necessary not only to use all the navigation and communication resources currently available in civil aviation but also to actively develop new architectural and software solutions ensuring the integration of navigation, communication, information, and computer resources, as well as their effective use.

This book is dedicated to solving these problems through the development of appropriate methods and algorithms.

Contents

Chapter 1
Analysis of the Problem of Ensuring the Noise Stability of the Civil Aircraft Navigation Systems

1.1 Formalization of the Unintentional Interference Space of the Civil Aircraft Navigation Systems

It should be noted that, despite all the advantages of using satellite-based radio navigation systems (SRNS) for the problem of navigational sighting of civil aircraft, there are so-called unstable navigation zones in the satellite navigation field. In some of such zones, interruptions in SRNS signals can reach one or even tens of minutes. The analysis showed that the cause of these phenomena was unintentional interference with the civil aircraft navigation systems. This interference can be of two main types [1]:

(1) in the form of harmonic components from radio facilities operating in adjacent radio wavebands;
(2) from radio facilities located within the navigation signal spectrum.

It appears possible to divide all unintentional interference with the civil aircraft navigation systems into three source-based groups [2]:

(1) natural,
(2) radio electronic, and
(3) associated with the flight of the aircraft (flight-related).

The first group includes interference associated with reflection from the underlying surface, mountains, and buildings. This interference is most meaningful during low-altitude flights between 400 and 1000 m. The danger of these reflections lies in the fact that such flights take place at those stages of the aircraft route, which require a limited time resource (landing approach). In addition, interference related to the specifics of flights in the polar latitudes can be attributed to this group. The lack of satellite technologies is most clearly seen in these areas, which is mainly determined by the fact that navigation satellites are very low above the horizon in the zone of the

© The Author(s), under exclusive license to Springer Nature Singapore Pte Ltd. 2021
D. A. Zatuchny et al., *Noise Resistance Enhancement in Aircraft Navigation and Connected Systems*, Springer Aerospace Technology,
https://doi.org/10.1007/978-981-16-0630-4_1

Far North, which leads to a sharp deterioration in the basic accuracy characteristic of SRNS-based aircraft positioning—a geometric factor.

Other interference that shall be attributed to the first group as natural interference also should be noted [3]:

(a)   Interference due to ionospheric effects. This effect is most unfavorable in the region of the geomagnetic equator and the auroral regions (near the magnetic poles). Inhomogeneities in the ionosphere are caused by solar activity and lead to daily changes in the passage of radio waves (day/night), and more adverse changes are caused by the 11-year solar cycle. Rapid changes in the ionosphere can lead to loss of the SRNS signal.

(b)   Interference due to atmospheric effects. It should be noted that the impact of these effects on the functioning of the civil aircraft navigation systems is relatively small. For example, intense precipitation leads to a weakening of SRNS signals only by fractions of a decibel. This fact does not have a special impact on the system operation. Tropospheric effects shall be taken into account already in the development of the system.

The second group shall include unintentional interference with the civil aircraft navigation systems associated with noise and industrial interference. These interferences most effectively affect small wavelength navigation signals, which corresponds to signals received from the navigation spacecraft (NS) transmitted in the decimeter range, as well as the transmission of information from civil aircraft board to the ground data receiving stations via VDL-4 or 1090 ES data lines in the automatic dependent surveillance mode (ADS).

Based on the analysis of statistics accumulated for the period of using SRNS for navigation support of civil aircraft, a conclusion can be made about the physical nature of the sources of such unintentional interference related to the onboard systems: leakage radiation or harmonics of VHF communication equipment, out-of-band and leakage radiation of satellite communication equipment, etc. Ground-based sources of such interference currently include mobile and fixed VHF communication equipment, radio links operating in the SRNS frequency band, harmonic components of radiation from television transmitting centers, some radar systems, and mobile satellite communication systems. It should be noted that the probability of such interference depends on the following factors: a set of state-formulated rules and recommendations in the field of spectrum use and frequency distribution, and level of compliance with these rules and recommendations in each state or region.

The third group includes unintentional interference related to the maneuver of the aircraft and, as a result, the shading of the NS signal by aircraft structural elements, as well as the aircraft landing on bare space.

There are two fundamental aspects that should be taken into account when assessing the threat arising during the civil aircraft flight associated with unintentional interference with civil aircraft navigation systems [4]:

(a)   probability of interruptions in SRNS work;
(b)   consequences of interruptions in SRNS work.

A detailed study of these aspects can determine the need to use methods to reduce the effect of unintentional interference on the civil aircraft navigation systems and to determine the level of threat associated with this interference. Such an analysis is necessary when flights are characterized by varying degrees of probability of serious consequences.

The probability of an SRNS interruption can be assessed based on operating experience. At the same time, the specifics of a particular civil flight region are to be taken into account. It should be noted that large cities with a significant number of sources of radio frequency interference, industrial areas, etc. are more susceptible to unintentional interference than sparsely populated areas. Thus, the probability of unintentional interference for industrial reasons in the European part of the Russian Federation is much higher than, for example, in Eastern Siberia.

It should be noted that ionospheric errors have a slight effect on the functioning of the civil aircraft navigation systems in the middle latitudes. However, in the equatorial region, and to a lesser extent in the near-polar latitudes, these errors can cause a loss (or poor-quality transmission) of a signal from one or several NSs. This situation can be compensated by the fact that there is a relatively large number of satellites observed in the equatorial region [5].

SRNS interruption is proposed to be considered not a complete system failure, which is an unlikely event, but a situation when the determined radio navigation parameters exceed certain tolerances, i.e. a failure to comply with the following set of inequalities:

$$\Delta x \leq \Delta x_{\text{доп}},\qquad(1.1)$$

$$t_{\text{пер}} \leq t_{\text{доп}},$$

where Δx—error caused by unintentional interference (interferences) when using SRNS, $\Delta x_{\text{доп}}$—maximum permissible error when using SRNS, $t_{\text{пер}}$ —SRNS interruption, and $t_{\text{доп}}$—permissible SRNS interruption.

Let us introduce the following notations:

$X = (x_1, \ldots, x_n)$—set of civil aircraft flight conditions,

$Y = (y_1, \ldots, x_m)$—set of regions differing in industrial production and underlying surface,

$Z = (z_1, \ldots, z_k)$—set of civil aircraft flight stages, and

$\Pi = (\Pi_1, \ldots, \Pi_l)$—set of unintentional interference with civil aircraft navigation systems.

It is assumed that an unintentional interference can introduce an error in the navigational sighting of the civil aircraft only with a certain population of elements from sets X, Y, and Z. If $D = \{X, Y, Z\}$, then it makes sense to introduce a relation $\Pi_i \to C_i$, where C_i—set of elements belonging to set D and conducive to an SRNS interruption due to unintentional interference Π_i.

Let's denote Q_i—an event of SRNS interruption due to unintentional interference Π_i. The probability of SRNS interruption due to unintentional interference Π_i is found as follows:

$$P(Q_i) = P(C_i)P(Q_i/C_i), \tag{1.2}$$

where $P(Q_i/C_i)$—probability of SRNS interruption due to Π_i with a certain set of factors.

When assessing the effects of interruptions in the civil aircraft navigation support provided by the SRNS, the following factors should be taken into account:

(a) Type of airspace. The limited time resource for corrective actions is more vital for the aerodrome area than for the airspace in which civil aircraft are flying on high-altitude routes.

(b) Air traffic intensity. For example, in areas with high air traffic intensity, the use of radar surveillance or the use of appropriate procedures by pilots may be impractical due to a workload.

(c) Level of service. For example, in the case of a flight through a zone where, for some reason, there is no SRNS-based navigation support, and when less-stringent requirements for the civil aircraft crew are presented, dead reckoning will be sufficient.

(d) Availability of other navigation systems. If it is possible to carry out navigational sighting using equipment based on civil aircraft navigational sighting using not only SRNS but also its integrated use (integration) with the inertial navigation system (INS), civil aircraft navigation systems will not be affected by SRNS interruptions.

(f) The scale of impact due to an interruption in navigation service is determined by the size of the geographic area, within which high-quality navigation support for the aircraft is impossible, and by the duration of this interruption.

(g) The ability of the civil aircraft crew and ground services to quickly assess the navigation service interruption threat level.

Table 1.1 provides a list of measures to improve the resistance of consumer onboard equipment to unintentional interference.

1.2   Analysis of the Noise Stability of the Civil Aircraft Navigation Systems Using the Example of the Sukhoi Superjet 100 Aircraft

According to the International Civil Aviation Organization (ICAO) standards "Aviation Telecommunication", Volume 1, Appendix 10, GPS SRNS signals or an integrated GLONASS SRNS and GPS signal or a GLONASS SRNS signal can be used.

Table 1.1 Potential measures to improve the resistance of consumer onboard equipment to unintentional interference

Item No	Interference immunity measures	Possible gain in relation to standard SRNS receivers, dB	Possible gain in relation to standard SRNS receivers, %	Notes
1	Improvement of the receiving antenna directional pattern at low elevation angles	12–14	27	Can be used in civil navigation systems
2	Control of directional pattern reducing sensitivity in the interferer direction	21–23	up to 85	For use, information on the direction of interference is required
3	Antenna array with signal polarization	11–13	up to 45	Not to be used under all operating conditions
4	Improved receiver signal processing	17–19	6–9	More research is needed to explore the possibilities of using
5	Integration of SRNS receiver with INS	12–14	25–35	Cost depends on INS level
6	Use of dual-frequency receivers L1, L2	7	25–40	
7	Use of triple-frequency receivers	9	37–45	

However, the following should be noted. With GPS flights introduced, interruptions in SRNS signals in some zones can reach one or even tens of minutes.

The revealed fact that GPS SRNS is subject to various kinds of interference causes its inapplicability for navigation problems in some cases. In contrast, GLONASS SRNS satisfies the requirements for noise stability. This is determined by the fact that GPS SRNS provides code division of signals, and GLONASS SRNS—frequency division of signals [6]. Thus, to degrade the quality of GPS SRNS, it is enough to put interference at the same frequency, and to achieve this result in the case of GLONASS SRNS—to put several interferences in the frequency range.

The characteristics of GLONASS and GPS SRNS, which affect the noise stability of the system to some extent, are presented in Table 1.2.

The advantages of GLONASS SRNS over GPS SRNS in terms of noise stability are quite difficult to use for the following reason. To date, civil aviation practically does not use domestic aircraft. Out of about a thousand aircraft operated by airlines operating flights to/from the Russian Federation, 7 Yak-42 airplanes, another

Table 1.2. Characteristics of GLONASS and GPS SRNS determining the noise stability of consumer receivers

Parameter	GLONASS	GPS
Signal division method	Frequency-based	Code-based
Signal carrier frequencies, MHz	1602.5625 to 1615.5 1246.4375 to 1256.5	1575.42 1227.6 to 1176.45
Signal band, MHz, standard precision	1.022 10.22	2.046 (C\A-code) 20.46 (P-code)
Repetition cycle, ms	1 ms	1 ms (C\A-code) 7 days (P-code)
Clock frequency, MHz	0.511	1.023 (C\A-code) 10.23 (P-code)
Signal level at L1, L2 receiver input, dBW	-165.5 to -157* -162	-164.5 (P-code) -163 (C\A-code) -166 (P-code) -166.8 (C\A-code)**
Signal polarization	Right-hand circular	Right-hand circular
Orbital altitude, km	19,100	20,145

*—at the output of an isotropic receiving antenna with linear polarization, i.e. with a loss of 3 dB
**—in the long term

3 airplanes developed at Gromov FRI according to the design of the Yakovlev Design Bureau (these aircraft have the ability to use GPS/GLONASS/GBAS signals), 1 aircraft equipped with all possible navigation receivers (the RF President board), and 2 aircraft of Special Flight Department "Rossiya" are used. The main civil aircraft (Boeing Airbus 320) uses a GPS signal.

There is an indigenous Russian aircraft—Sukhoi Superjet 100 shown in Fig. 1.1. But this aircraft is also equipped with old foreign avionics by Thales, which only confirms the existence of the problem outlined earlier.

It has been affirmed that it is necessary to be able to receive signals from two constellations on board, but proposals for real sharing of GLONASS and GPS signals, although there is such an opportunity in the ICAO Standards, elicited no response from representatives of foreign aviation services.

The flight-navigation equipment of the Sukhoi Superjet 100 aircraft.
provides the crew with the data necessary for the flight.
The flight-navigation equipment includes the following systems:

– systems and instruments for measuring flight parameters,
– instruments for measuring the spatial position and direction of flight,
– systems and instruments for landing and taxiing,
– non-cooperative flight-navigation systems,

Fig. 1.1 Aircraft "Sukhoi Superjet 100"

– cooperative flight-navigation systems, and
– navigation computers.

We will analyze the problem of meeting the noise stability requirements taking into account the fact that modern civil aircraft are almost fully capable of navigating with the use of consumer navigation equipment (CNE) operating on a signal received from navigation spacecraft (NS) included in SRNS. Also, many CNE examples are capable of locating the civil aircraft on the basis of SRNS and INS integration (integrated use) [7].

The GLONASS global navigation satellite system is a satellite navigation radio system that provides an approach and a continuous determination of the aircraft's navigation parameters using satellite navigation system signals, as well as information received from onboard equipment (navigation data), ground-based augmentation system, and augmentation satellite systems.

The global navigation satellite system includes the following equipment: GLONASS receiver–computer unit, GLONASS amplifier, and GLONASS antenna.

The Global Positioning System (GPS) is an auxiliary radio navigation system that provides the crew and FMS computers with navigation information about the position, heading, and speed of the aircraft. To this end, the system uses signals from the satellite constellation. The calculated navigation information, as well as the number of visible satellites and the integrity of the system, is displayed on the FMS computer consoles. The system includes two antennas that receive radio signals from satellites. The signals received are sent to multi-mode receivers of the instrument landing system.

The inertial system installed on the Superjet 100 civil aircraft consists of three independent channels. Each channel includes an inertial computer and an inertial computer configuration module. The inertial system provides the crew and other aircraft systems with information about its spatial position (pitch angle, roll angle, and yaw angle), heading, ground speed, and current location. The values of angular velocity and linear acceleration of the aircraft are calculated by the inertial computer and sent to the automatic flight control system, the flight data recording system, the flight management system, and the central computer system. The calculated information is displayed on the display.

When calculating the spatial position and flight direction parameters, the inertial system interacts with aircraft systems and receives data directly from sensors or computers. Depending on the coordinate system used, the inertial system provides the following data to the aircraft systems:

(1)   In the body-fixed coordinate system:

– longitudinal, transverse, and normal accelerations;
– roll, pitch, and yaw angular velocity.

(2)   In the local coordinate system:

– pitch and roll;
– angular pitch and roll rate;
– inclination angle of the flight path inclination angle and linear acceleration along the flight path;
– baroinertial vertical velocity and inertial vertical acceleration;
– platform azimuth.

(3)   In the Earth-fixed coordinate system:

– latitude and longitude;
– ground speed, north and east components of ground speed;
– baroinertial altitude;
– true and magnetic heading;
– true and magnetic track angle;
– angular track angle rate;
– true velocity and true wind direction;
– drift angle;
– acceleration along the track and perpendicular to the track;
– acceleration along the heading and perpendicular to the heading.

The inertial system has the following modes of operation:

– activation,
– stationary alignment,
– navigation,
– spatial position (backup mode), and
– flight termination.

The inertial computer consists of a chassis and a front casing with the three main components inside: inertial sensor unit, secondary power supply with electromagnetic interference and transient protection board, computer board and ARINC I/O interface.

The inertial sensor unit consists of three digital laser gyroscopes and three compensating accelerometers; each of the components is equipped with its own electronics, which allows it to be isolated from faults in adjacent sensors.

The inertial calculator uses the Q-FLEX QA950 accelerometer by Honeywell. Built-in electronics generate an output current proportional to the acceleration, which provides a measurement of both static and dynamic accelerations. The technologies of existing Honeywell products that have been certified by the US Federal Aviation Agency and the European Certification Center are applied in the computer and the ARINC I/O board. The microprocessor and application-specific integrated circuits (ASIC) have been tested in other commercial applications. The software is certified in accordance with DO-178B, level A. The computer and ARINC I/O interface board are designed with the possibility of expanding the memory capacity and throughput rate.

The inertial computer configuration module (ARM) is a non-volatile memory device that stores configuration data of the ARM inertial system [8].

"Activation" mode.

After feeding the power supply from the aircraft onboard network and activating the inertial system, it enters the "Activation" mode, which lasts approximately five seconds. During this time, the built-in monitoring tools perform a function check that can detect over 90% of the failures, and the configuration data is loaded from the ARM memory.

"Stationary alignment" mode.

If there are no failures, the INS switches to the "Stationary alignment" mode, during which the INS determines the current position of the aircraft taking into account the latitude/longitude coordinates received from GPS or entered by the crew. The spatial position mode is activated simultaneously with the stationary alignment mode. Thus, the indication of the spatial position parameters (roll, pitch, etc.) is provided immediately after the INS is turned on, but before it switches to the navigation mode.

The duration of the "Stationary alignment" mode varies from 3 to 20 min, and depends on the current latitude and type of alignment selected: standard or fast (to be set in the ARM-stored configuration file of the aircraft). During the alignment, the INS calculates the latitude coordinate and the North line after aligning the INS axes with the local vertical according to the components of the Earth's rotational speed corresponding to each axis. If the aircraft moves during the alignment, the alignment is terminated. Re-switching to the stationary alignment mode occurs 30 s after the end of the movement, and the whole alignment process begins anew.

If the latitude and longitude coordinates were entered by the crew, then, after the alignment is completed, the INS conducts a compliance check of the entered values and the last memory-stored values measured before the previous power-off. If the difference exceeds 1°, the crew receives a message about the need to re-enter the coordinates. If the coordinates re-entered also do not coincide with those stored in the INS memory, but differ by less than 0.1° from the ones entered for the first time, they are accepted. A compliance check is not carried out for the entered values if the INS worked in the spatial position mode before power-off or the "Ground service" flag was set in the INS memory.

Upon alignment termination, the INS conducts alignment correctness verification; for this, the values of the cosine and sine of the given latitude (obtained from GPS or entered by the crew) and the INS-calculated latitude are compared. If the difference in values does not exceed 0.01234 for cosine and 0.15 for sine, and the compliance check for the coordinates entered by the crew is successfully completed, then the INS switches to the navigation mode. Otherwise, the INS issues a message about the need to re-enter the coordinates.

Instruments measuring the spatial position and direction of flight of the aircraft are designed to calculate the spatial position and navigation parameters of the civil aircraft for the unambiguous perception of the air situation, civil aircraft position, and flight mode by the crew, as well as to work with automatic flight control equipment.

Parameters calculated by spatial position and flight direction instruments:

- current spatial position,
- ground speed,
- heading angle,
- current values of wind direction and force,
- drift angle,
- track angle error,
- across-track displacement,
- design heading,
- values of current navigation characteristics,
- required navigation characteristics in accordance with the flight mode,
- inertial vertical velocity, and
- magnetic and true heading.

On the basis of the parameters listed, the spatial position and flight direction instrument system calculates the spatial position of the aircraft in accordance with the 3D Earth-positioning coordinate system (WGS-84) (Fig. 1.2).

"Navigation" mode.

If the alignment correctness verification and the verification of coordinates entered (if they were entered by the crew) are successful, the INS switches to the "Navigation" mode and remains there until the discrete input is opened or the power supply is interrupted.

"Spatial position" mode.

The spatial position mode is a backup mode, which provides minimal initialization of the inertial system data during the flight and is automatically activated after a short power supply failure of the inertial system.

"Flight termination" mode.

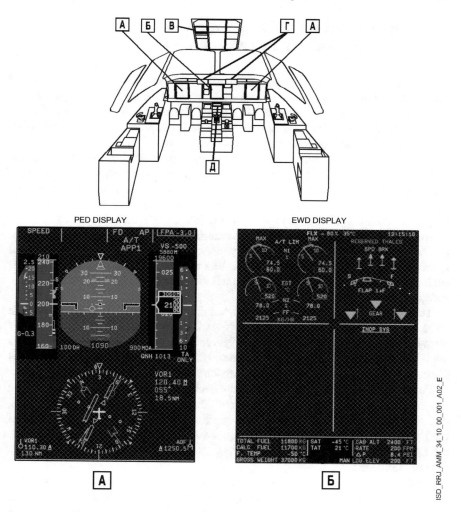

Fig. 1.2 Controls and indicators

In the first 5 s of the "Flight termination" mode, INS records and saves the following information in its internal non-volatile memory: navigation quality data, service data (in accumulation mode), auto-calibration data, and other parameters.

Based on the data provided in this section, we can draw the following conclusions:

(1) The noise stability of the GLONASS system is higher than that of the GPS system.

(2) The use of foreign avionics on board the civil aircraft including

"Sukhoi Superjet 100" does not enable taking advantage of these benefits [9–11].

1.3   Mechanisms for the Use of Expert Systems in the Design of Aircraft Airborne Systems

This section sets the task of forming an approach that involves consideration of specific uncertainties arising from the use of the results of the development of civil aircraft flight-navigation systems (FNS) in terms of ensuring noise stability. A feature of the proposed mechanism for organizing such developments is the use of expert assessments as important components of soft computing [12].

According to the proposed methodology, an expert description of the design outputs creates conditions for project management based on qualitative assessments, just as is done in the project management and scheduling theory. We will assume that the basis for compensating for specific uncertainties in the civil aircraft FNS design outputs is the assessment of the operational reliability of the FNSs created.

An important feature is the absence of criteria that characterize the importance of parameters determining the sustainability of the results of individual works, as well as the reliability assessment of the project as a whole. Therefore, to compensate for such uncertainties, the actual design of civil aircraft FNSs will be considered as a complex system.

The term "complex" system will be used not in contrast to "not complex", but as its characteristic. Since the design involves the operation of facilities with fuzzy characteristics of an indefinite number of parameters, to describe a complex system we will not use the parameters of the interrelations between the project stages, but the characteristics of the functions that implement these interrelations.

For example, we will consider the design technology as a complex system taking into account the impact of the "human factor", the impact of uncertain natural factors, etc. A special feature of the proposed approach is not only the development of qualitative methods for evaluating the effectiveness of the FNS project, but also the optimization of aircraft operation taking into account indicators of resistance to external influences [13].

Traditionally, the design completeness has to be evaluated using the data obtained from an experiment with the probability of work success calculated by the Monte Carlo method. We will assume that the stability of FNS work is characterized by assessments of the functional characteristics defined in the design specification provided that the previous works ensuring the commencement of operation have been completed. In this regard, in contrast to the structural reliability determined by time to failure, recoverability, and maintainability of FNS, it is proposed to consider functional reliability determined by the performance of regular functions under the condition of the dynamics of external influences.

In this paper, for the algorithmization of the examination of the acceptability of design results (as a complex system), the task is to form a system that allows ensuring the functional reliability of developments, i.e. the reliability of obtaining a regular result of the use of FNS in a given time interval, under given conditions.

The analysis of FNS work showed that causes of dangerous failures also include incorrect requirements: to limitations in the achievement of the development target function, to hardware and/or software; to reliability (lack of development of the necessary functions to ensure the stability of work in various modes); in addition: accidental failure of mechanical devices; software errors; disturbances in the power supply system (power failure, voltage reduction, power switching); environmental influences (electromagnetic, thermal or mechanical phenomena); operator errors, etc. This requires an expert solution to the two problems: introduction of the stability criterion of the designed FNS and identification of factors affecting the performance of its regular functions.

The considered problem of taking into account the specific uncertainties arising from the use of the design outputs of the civil aircraft FNS requires the analysis of external influences on the results of achieving the process design objectives, as well as the service life of the aircraft. Estimates of such uncertainties are commonly referred to as "risk". In this paper, we quantify the risk using the product of the probability of a risk event by the amount of damage from its occurrence.

The significance of the impact of uncertainties on the sustainability of the FNS creation results confirms the relevance of the risk management task at each stage of the civil aircraft life cycle, from design to removal from service. In this consideration, risk management is proposed to be conducted by identifying threats, setting countermeasures, and analyzing the results of compensation for external influences. For this purpose, the problem to establish whether the changes in risk correspond to the standard FNS reliability values is set.

One of the risk management models based on the relationships between threats, countermeasures, and risks, given in GOST R ISO/IEC 15,408-1, is shown in Fig. 1.3.

GOST R ISO/IEC 15,408-1-2002 "Information Technology—Security techniques—Evaluation Criteria for IT Security" defines the "General Criteria" (GC) designed to assess the safety performance of IT tools and systems. By establishing a common base of criteria, GC make FNS functional stability assessments significant for the formation of expert conclusions on the sustainability of the design results. The completeness of these conclusions is achieved through the analysis of the fulfillment of the general set of requirements for the safety functions of the civil aircraft tools and systems, as well as in accordance with the degrees of confidence in the results of such an analysis [14].

According to the formulation of the problem being solved, a more detailed consideration of the specifics of ensuring the FNS design sustainability determines the need to apply expert assessments of the risks of their work under external influences on the aircraft. Due to the fact that the aircraft is considered as a man–machine system, it is proposed to characterize it as an object of protection dividing into physical, logical, and personnel components [15]:

(a)   *Physical objects* include material components or a group of components within the aircraft. Such objects may include FNS technical means, transmitting and receiving equipment, special tools installed on the aircraft, etc.

(b)   *Logical objects* are informational in nature. They may include process data, control algorithms, and specialized information that ensures the functional

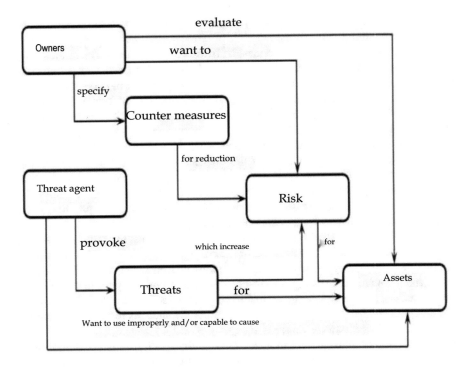

Fig. 1.3 Risk management model according to GOST R ISO/IEC 15,408-1

performance of the FNS. Indirectly, objects of this type may include indicators, whose reduction affects the FNS operational and production efficiency.

Civil aircraft control processes are a special form of logical objects. They contain the automation logic used in the respective systems. Failures of FNS and computer systems may occur both at the physical level (for example, destruction of a memory element as a result of electrophysical effects) and at the logical level (for example, change in the contents of a memory cell). Such effects can lead to violations of the control signal integrity or the stability of the very FNS operation process.

(c) *Personnel objects* include professionals characterized by knowledge and skills relevant to their production activities. For example, failures of FNS and computer complexes can significantly affect the functioning of the civil aircraft as a whole, despite the integrity of physical and logical components.

Thus, it can be assumed that one of the central problems in the technology of maintaining the sustainability of the FNS design outputs under the influence of external factors is minimizing the risks of faults or failures of each of the components of the protected object. An important feature in solving such a problem is the frequent inconsistency between the FNS reliability requirements defined by their design specification and the possibilities of design solutions proposed by developers. In this case, the optimization of the solution to the inverse reliability calculation problem is required, which implies the development of recommendations for the refinement of the FNS design based on the provisions of their design specifications for this aircraft.

An essential characteristic of inverse reliability calculation problems is the fundamental incorrectness of their formulation due to the multivariance of possible solutions. As applied to FNS, the uncertainty of the formation of proposals for the design refinement as a result of the lack of rigidity in determining the parameters of the influence of external factors on the civil aircraft during flight is added.

The uniqueness of the onboard equipment requires the strict consideration of its characteristics. However, in practice, there is often a spread in their values due to deviations from the required manufacturing technology. At the same time, resources for the implementation of improvements are always limited [16].

Thus, it requires the introduction of expert recommendations for the introduction of design improvements, minimizing the uncertainty in FNS works in relation to the requirements of their design specification, while minimizing the resources spent, i.e. optimization of the expert solution to the inverse CA reliability calculation problem is required.

In this section, uncertainty refers to the characteristic of a situation in which the estimates of decision-making constraints are described by random processes. In our consideration, the following uncertainties are to be identified:

- stochastic (there is information available about the distribution of the failure probability over a set of solution outputs);
- operational (there is information available about the FNS work results, but there is no information about external influences);
- natural (there is information available about possible external influences, and there is no data on the results of these impacts with the proposed design solutions);
- a priori (there is no information about both external influences and the results of the proposed design solutions).

It is proposed to reduce the problem of justifying solutions under the conditions of uncertainty of these types, all but a priori, to narrowing the initial set of alternatives based on the information provided to the decision maker. Due to the need to take expert decisions, we will assume that the quality of recommendations increases under conditions of stochastic uncertainty when taking into account such characteristics of the decision maker's personality, such as attitude to the outputs of his/her decisions and risk appetite.

It is possible to use automated decision support systems under the conditions of a priori uncertainty in predicting the CA FNS stability. One of the promising directions in the construction of such systems is the use of adaptive control algorithms [17].

Practice shows that traditional prediction methods work if the interaction process between functional blocks is stationary or its characteristics change little in time. In this paper, design requirements for the FSN, which ensures the efficiency of response to external influences in case of uncertainty of the processes affecting functional blocks, determined the need to consider the value of the protected object taking into account the efficiency of interaction between the blocks of the FNS being created.

In our consideration, it is proposed to take the fault or failure risk assessment as the measure of the value of CA FNS components. Recall that risk assessment is considered as the product of the risk event probability by the amount of damage caused as a result of its occurrence. In this case, the damage may be direct or indirect [18]:

(a)   direct damage: reflects the cost of replacing the failed object;
(b)   indirect damage: reflects the damage caused by the failure of the object for which the designer or manufacturer is responsible. For example, such damage may include losses associated with process downtime caused by object failure. Indirect damage in relation to physical objects usually includes the consequences of an aircraft flight mission failure. Indirect damage in relation to logical objects often includes a loss of confidence in a designer or manufacturer, a loss of competitive advantage once achieved through the realization of intellectual property (for example, a groundbreaking technology), etc.

Risk assessment involves the step-by-step analysis of all functional blocks involved, starting with the ones immediately accessible to the threat, and further to the less accessible ones. We will assume that the risk assessment scheme consists of the following stages:

(1)   assessment of the initial risk;
(2)   implementation of risk mitigation countermeasures;
(3)   assessment of the residual risk.

If necessary, the second and third stages may be repeated, which is one of the ways to reduce the residual risk to an acceptable level. In particular, the second stage may include an assessment of existing measures for protecting and implementing plans to add corrective or additional countermeasures.

As the basic risk analysis stage, we will take a study of vulnerabilities in the aircraft design solutions, i.e. identification and characterization of weak points in the FNS components from the point of view of external influences. Such vulnerabilities may be forced into the design of the control system or be random, that is, they may be the result of the FNS design customer's misunderstanding of the functional environment features.

In this consideration, vulnerabilities are not limited to electronic components. Similar to the regularities of changes in the reliability of man–machine systems, the vulnerabilities of a civil aircraft can also increase if the characteristics of the environment change, if the components of the protected object fail, if such components cannot be replaced, etc.

Impacts that may occur in relation to civil aircraft will be considered as threats. In the practice of operating FNS, they appear in various forms, but the most common forms of external influences are as follows:

(a)   accidental natural: radiation and solar flares. These factors are poorly predictable and modelable.

((b)   predictable natural: ionizing electromagnetic radiation, geomagnetic distur-
       bances of the Earth's magnetic field, geomagnetic storms, and thunderstorm
       fronts.
(c)   unintentional artificial influence.

The current regulatory documents that determine the maximum estimates of
acceptable risks are very non-specific. Therefore, we assume in this section that
a list of objects or scenarios with estimates of the probability of risk events and a
gradation of the importance of their consequences shall be formed for each type of
aircraft and each flight mission. This list shall be a database for qualitative analysis
of actual risk. Based on the information from such lists, it can be assumed that the
reliability of personnel protection objects is largely determined by the management's
ability to identify measures that influence physical and logical components [19].

In accordance with the specifics of their activities, some developers accept relatively high levels of risk (in particular, research organizations), and some are, in
fact, conservative and refuse risks. Therefore, an assessment of the risk level of
the same man–machine system may be acceptable for one organization and unacceptable for another. Even individual units within the same organization may have
different opinions regarding risks or their acceptability. The responsibility of decision
makers implies the need for the most objective assessment of risks and their acceptability to prepare a response to risk events. Depending on the circumstances, decision
makers may resort to one or another combination of risk-influencing measures in each
situation. For example [20],

(a)   *elimination of risk from the design:* one form of risk mitigation is to amend the
       FNS design so as to eliminate this risk. The risk can be mitigated by completely
       canceling an unnecessary function or blocking access to it. Such an influencing
       measure may provide for conscious refusal from specific positions in the flight
       mission;
(b)   *risk reduction:* risk can be reduced to an acceptable level by taking counter-
       measures that reduce the likelihood of external influences or limit the conse-
       quences of a risk event. In this case, the principal limitation is the possibility of
       achieving an acceptable level of risk on the understanding that the risk cannot
       be completely eliminated;
(c)   *transfer or distribution of risk:* it is possible to enter into any insurance contract
       or agreement on transferring all or part of the risk to a third party. A typical
       example is a subcontract for certain operations or services. Such a solution may
       not always be effective, since it may not apply to all aircraft components. An
       insurance contract may compensate for some types of damage, but not damage
       related to logical components, for example, loss of confidence from consumers
       of FNS functional blocks;
(d)   *elimination or revision of ineffective protective measures:* within the framework
       of minimizing the effects of risk events, protection measures used shall be
       identified and harmonized so as to focus on the improvement of the most
       effective and efficient measures; and

(e) *risk acceptance*: there is always an option to accept the risk considering it as
part of the aircraft operating costs. Developers shall take some risks if they
cannot be mitigated or transferred at optimal costs.

In these cases, the following solutions can be used as countermeasures taken to
reduce the risk of external influences to an acceptable level:

– redundancy of information channels without changing the operational characteristics of the FNS functional blocks;
– algorithmization of control in the man–machine system assuming that the decision
  maker chooses a command, for example, "sleep"—a temporary interruption of
  the active functioning of a specific FNS block, "key"—switching to a redundant
  channel, set, power source, and "shift"—implementation of aircraft maneuver;
– automation of control in the technical system involves issuing commands to
  activate the actuators of the external influences protection system;
– automation of control in the emergency system—involves emergency shutdown
  of the functional block and, if necessary, its isolation.

According to the objective of this section to ensure the sustainability of the CA
FNS design outputs, we will consider the characteristics of the expert system (ES)
allowing to ensure the functional reliability of complex human–machine systems. In
our case, we will assume that the main functions of such an ES can be divided into
three main groups:

– formation of an expert group;
– ensuring the interaction of experts within the group;
– development and adoption of a technical solution to the target problem.

To date, many international organizations have the most advanced interface with
control functions in the ES. As a reference system, we can call the expert management system adopted by the International Electrotechnical Commission (IEC). The
standard expert management functions in such a system are the following:

– formation of an expert group (*expert management system*);
– ensuring the interaction of experts within the group (*meeting registration system
  (MRS), web conferencing, etc.*);
– development and adoption of a technical solution to the target problem (*collabo-
  ration tool, electronic vote, and comment*).

The work of such an expert management system is based on an expanded set
of functions: for example, access to a database and reference information about
previously developed documents (*library server, standard in database format, and
electropedia*); and system administration (*technical server, management server, etc.*).
The expert is provided with documents currently used, list of committees (expert
groups) he/she being a member of, available standard set of actions (delete, create,
notify, etc.) for working with documents.

An important advantage of using such a computer system is the virtualization
of ES collections. The video conference function allows organizing the interaction

of experts in real time without the need for their presence in one place. The expert administration system allows for the registration of experts appointed to participate in the ES work. This model allows monitoring all the changes associated with the personal data of experts and coordinating their work.

To organize documentary support for the ES Decision Support (ESDS), this paper proposes mechanisms that are determined by the features of managing the reduction of risks associated with the influence of the external environment on the sustainability of the FNS design outputs. This form of documentary support involves a set of functions of planning and implementation of the three stages of information processing [21]:

– operational problem-oriented collection of documents that are integrated from various external sources;
– documented multi-user analysis;
– formation of corporate documentary reporting.

The specificity of the problem we are considering determines that the documentary data received in the ESDS is characterized by a high scaling level and complexity of multi-level indexing, using their semantic interpretation, with transaction complexity and other features that require thorough verification, maintenance and support of relevance, timeliness, and reliability. Therefore, the proposed model offers a software service for working with large volumes of documentary information, which provides the following solutions:

1.   Decomposition approach focused on the document flow administration of small and large networks. In this case, it is possible for single developers to plan offline business processes and update personal digital libraries.
2.   As a formal model of an ESDS information infrastructure, it is proposed to use a "cataloged system", the elements of which are object-oriented, process-oriented, event-oriented catalogs, registers, directories, and repositories of multidimensional monitoring data that will be able to sufficiently take into account the specificity of the application area.

After registration, information about competencies (input data is presented in Fig. 1.5) is entered for each participant (Fig. 1.4). The executive roles of the network SDS participants are distributed according to the results of the competency analysis. At the same time, documentation done in a systematic and consistent manner will allow the ESDS to receive the status of a formal organizational system.

An important aspect of ensuring the functional reliability of the ESDS is the algorithmization of building information and communication policies taking into account the distribution of expert functions in accordance with their competence levels. One of the solutions to this problem is shown in Fig. 1.6. Data for entering information presented in this form allow navigating through the nomenclature list of documents, thus providing an opportunity to select the performer of a particular business transaction. In this paper, for example, such an operation is to ensure the sustainability of the CA FNS design outputs during external electrophysical influences (EEI).

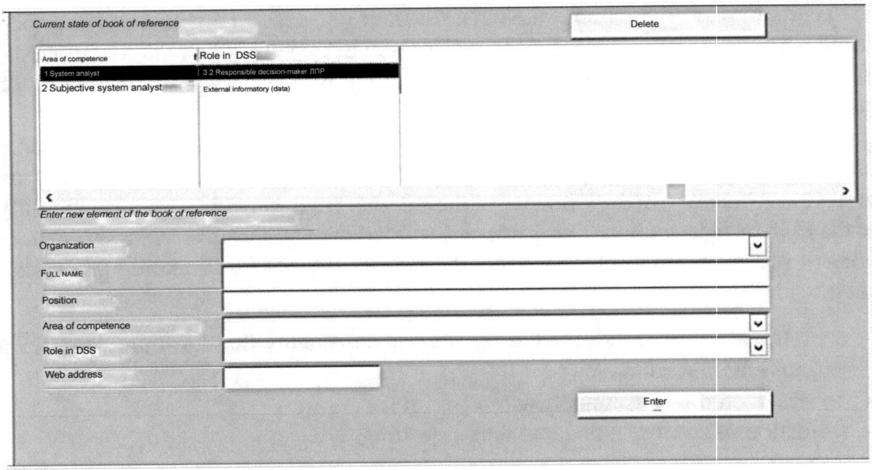

Fig. 1.4   ESDS participant registration screen

Code	Reference
1	System analyst
1.1	Problem identification. Intersubjective statement
1.2	Subjective problem statement
1.3	Tasks definition of expert groups
1.3.1	Decomposition of initial task and description of initial sub-tasks for expert groups
1.3.2	Determination of composition and sequence of applying root algorithm (solving graph)
1.3.3	Determination of temporary and other regulations
1.3.4	Correction of sub-tasks formulation at their consecutive solving
2	Subjective system analyst
3	System analyst for document support
4	Expert for supporting online database by EI
4.1	Operator of electronic library
4.2	Operator of problem-oriented analysis of telemetric data
4.3	Operator for support of online database on incidents
4.4	Operator for support of online database on protection equipment
4.5	Operator for support of online database on products lifecycle
4.6	Operator for support of online database on field components
5	Expert for supporting online database by EI
5.1	Operator for configuration of information sources
5.2	Operator for problem-oriented setup of intellectual functions and means of expert analysis

Fig. 1.5   Data to enter information about the competencies of the network SDS participants

This service allows the network document management administrator to program the procedures of network expertise process management, distribution of responsibilities, and organizational management of the generation of corporate reporting documents.

It should be noted that, in contrast to the classical ES building systems, the proposed model of ESDS supports the indexing of fragments of the CA FNS information infrastructure, which makes it possible to use the entered structuring schemes not only for human–machine (interactive) but also automatic (software) navigation

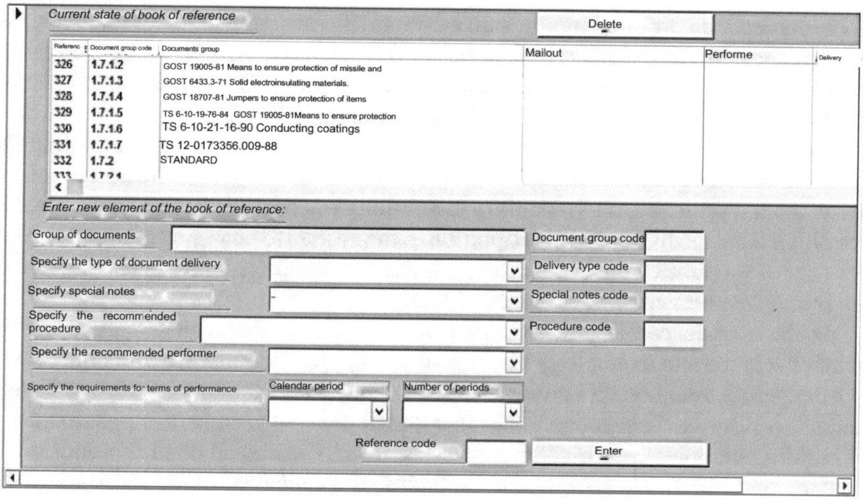

Fig. 1.6 Entering the regulations of network business processes

in the information field of directories, catalogs, registries, etc. in the interests of automated applications, services, and intellectual functions [18, 19].

Consideration of the features of using information technologies in ensuring the stability of CA FNS work under external influences has shown some possibilities of expert support for the management of the FNS development. In the proposed methodology of information support for forecasting the aircraft response to external influences, an attempt was made to combine methods of computer data processing and expert analysis of the operation of functional modules. Such an approach shall ensure the effectiveness of decision-making based on improving the accuracy and speed of data processing, and, accordingly, choosing the scenario for adapting the overall work of the civil aircraft to external influences taking into account the preferences of decision makers.

System consideration of the aircraft creation and work organization technologies showed that the semantics of interfaces largely determines the stability of the interaction between the FNS functional blocks and their relationships with the human operator, and pragmatic aspects affect the dynamics of the FNS improvement. At the same time, the friendliness of the semantics of interfaces in the form of positive feedback often entails the improvement of pragmatic components, which, in turn, can determine the conditions for improving the semantics of interactions, etc.

Application of the conditions of thermodynamic equilibrium in a complex system to the problem under consideration allowed formulating the principles of ensuring the homeostasis of the functional characteristics of CA FNS [14]:

The first principle: harmonization of the mutual influence of the results of the pragmatic and semantic aspects of interactions between FNS elements. Harmony refers to the isomorphic balance of the value of information interactions between

FNS elements determined by the sum of pragmatic and semantic components. An example of harmonization is the equality of the rate of adaptation of the passage of information signals between FNS elements and the rate of the operator's adaptation to the perception of data under changing external conditions.

The second principle: formation of the target-based focus in improving the semantic aspects of FNS work in a changing external environment. As a result of the use of such a principle, the possibility of fulfilling the complex system survivability condition is formed: the time of adaptation of the system to changing conditions shall not exceed the time of the onset of irreversible changes in its work occurring due to the influence of external environmental factors.

As the experience of operating domestic and foreign FNSs shows, the development of science and technology in terms of solving problems of reducing the causes of a dangerous situation on a civil aircraft under the complex effect of external factors remains a priority. This is due to the fact that at present there is no single ideology of building complex information processing for CA technical condition indication systems and efficiently solving the problems of intellectual assessment and forecasting of the development of a dangerous situation during the degradation of FNS characteristics.

When applying existing methods, the well-known assessments and forecasts are the results of complex processing of strongly averaged-in-size parameters and time intervals of some generalized external influence and other factors manifested in the respective operational conditions of the aircraft.

The practical value of the assessments and forecasts under consideration for taking timely measures to prevent the occurrence of a dangerous situation on the aircraft caused by external influences remains very low at the present time. These assessments and forecasts are suitable mainly for analyzing the causes of the unusual events that have already occurred on the aircraft in the course of work of the operational group during the aircraft operation and in the course of work of the flight development test commission. In real time, such assessments and forecasts allow to reliably assess and predict only the "strongest" effects of degradation of characteristics, such as those associated with "solar flares", passage of civil aircraft in known areas of enhanced radiation, and long-term operation.

A more detailed specification of the assessments and forecasts of the causes of a dangerous situation on the aircraft and the parameters causing its effects is associated with the high complexity of modeling the processes under consideration and the changing availability of data for this modeling during the aircraft operation. At the same time, the most effective method for a particular operation stage is a specific assessment and forecasting method. The selection of such a method from a set of possible methods is associated with the introduction of certain rules and is a smart solution. As the analysis of practical experience gained from the available information sources for assessing the FNS reliability shows, expert systems, which are functional blocks of the Industrial Internet (IIoT), can advance the use of the proposed solutions to improve the complex system design processes.

References

1. Aeronautical Telecommunications Annex 10 to the Convention on International Civil Aviation—Montreal, ICAO, Sixth Revision (2006)
2. Avtin IV (2017) On remote determination of complex dielectric permeability using radiopolarimetry methods. Collected book: VII All-Russian armandovskiye readings. Current problems of remote sounding, radio detection and ranging, wave propagation and diffraction. Materials of the All-Russian scientific conference, pp 384–386
3. Avtin IV (2017) On building-up a polarization image of a radar target. Collected book: VII All-Russian armandovskiye readings. Current problems of remote sounding, radio detection and ranging, wave propagation and diffraction. Materials of the All-Russian scientific conference, pp 387–391
4. Avtin IV, Trushin AV (2015) On detection and resolution of low-contrast low Doppler radar targets within the resolution cell. Nauchniy Vestnik of the Moscow State Technical University of Civil Aviation 222(12):80–84
5. Akinshin NS, Bystrov RP, Menshikov VL, Potapov AA (2018) Specifics and methods of avionic equipment jam resistance improving. Uspekhi sovremennoy radioelektroniki 2, 3–21
6. Akinshin RN, Zatuchnyy DA, Shevchenko DV (2016) Reducing the multi-path propagation impact when transmitting data from the aircraft. Informatizaciya i svyaz 3, 6–12
7. Akinshin NS (2001) Data protection in avionic data transmission and processing systems. In: Conference proceedings "Voprosy radiotekhniki i radiolokacii" TulSU, TAII, Tula
8. Didenko NI, Eliseyev BP, Sauta OI, Shatrakov AY, Yushkov AV (2017) Radio-technical support of civil and military aviation flights: strategic problem of the arctic region of Russia. Nauchniy Vestnik of MSTUCA 20(5):8–19
9. Zatuchnyi DA (2018) Analysis of various jamming impact on the civil aircraft navigation systems. Informatizaciya i svyaz 2, 7–11
10. Zatuchnyi DA (2012) Reducing errors in data transmission from the aircraft in mountainous areas for the VHF band by improving the transfer point selection. Nauchniy Vestnik of MSTUCA 176, 150–153
11. Zatuchnyy DA, Logvin AI, Nechayev EE (2012) Problems of ADS mode implementation in the Russian Federation. Printing and publication department MSTU CA
12. Kozlov AI, Sergeyev VG (1998) Radio waves propagation along natural paths. MSTUCA
13. Kozlov AI, Garanin SA (2005) On the development of mathematical simulations reflecting impact of radio jamming and random effects affecting the aircraft on the navigation parameters to be determined. Nauchniy Vestnik of MSTUCA, series Radiofizika i radiotekhnika, 93
14. Romancheva NI (2018) Inverse problem of complex systems elements quality evaluation. In: Proceedings of the international symposium "Nadyozhnost i kachestvo", Penza, vol 1, pp 71–75
15. Romancheva NI, Yurkevich EV, Kryukova LN (2018) Mechanisms for development of technical documentation for the means of maintaining the aircraft survivability. Nauchniy vestnik of Saint Petersburg state technical university of civil aviation 3(20):39–48
16. Romanchev IV, Bannov VY, Romancheva NI, Yurkov NK (2008) Evaluation and anticipation of software reliability. In: Proceedings of the international symposium "Nadyozhnost i kachestvo", Penza, vol 2, pp 41–44
17. Romanchev IV, Yurkov NK, Romancheva NI, Bannov VY, Trusov VA (2006) Statistical analysis and diagnosis of malfunctional elements of the object. In: Proceedings of the international symposium "Nadyozhnost i kachestvo", Penza, vol 2, pp 44–45
18. Manual of air traffic services data link applications (1999) ICAO, Second revision
19. Required navigation performance (RNP) Manual (2008) Third revision
20. Severcev NA, Beckov AV (2009) System analysis of the theory of security. M.: Lomonosov Moscow State University
21. Yurkevich EV (2007) Introduction to the theory of information systems. M.: LLC Izdatelskiy dom Tekhnologii, 272 p

Chapter 2
Mathematical Models of Information Security in Civil Aircraft Flight-Navigation and Computing Systems

The problem of ensuring the necessary level of information security under the conditions of an ever-expanding list of threats, means, and methods for their implementation is very complex, requiring for its solution not just the implementation of a certain set of scientific, R&D, technical and organizational measures but also the use of specific tools and methods of data security to create their entire system. At the same time, ensuring the information security in modern flight navigation systems (FNS) and computing systems (CS) shall be not a one-time action, but a continuous process purposefully carried out throughout the creation and operation of the system with the complex application of all available tools, methods, and measures.

The most effective approach to solving the problem is redundancy. However, due to the large time input for recovering destroyed information and the large material costs of storing the recovery backup, it is necessary to determine how and in what cases it is advisable to use a computational resource of information [1].

2.1 Development of Models and Methods of Structural and Technological Redundancy of Software and Information Support of Civil Aircraft Flight-Navigation and Computing Systems

2.1.1 General Mathematical Model of Structural and Technological Redundancy of Software and Information Support of LAN-Based Flight-Navigation and Computing Systems

The design of the optimal logical structure of software and information support of the LAN-based civil aviation (CA) FNS and CS is carried out on the basis of a

© The Author(s), under exclusive license to Springer Nature Singapore Pte Ltd. 2021
D. A. Zatuchny et al., *Noise Resistance Enhancement in Aircraft Navigation and Connected Systems*, Springer Aerospace Technology,
https://doi.org/10.1007/978-981-16-0630-4_2

structural approach, involving a complex procedure for "fitting" the problem-solving technology to the architecture of a specific or hypothetical local area network.

The structural approach is founded on the methods for designing the logic-level components of software and information support based on the use of a set of procedures for the sequential transformation of matrix and graph models of the canonical structure of the computing system into the logical structure of the computing system. This approach allows to formalize, algorithmize, and computerize the problem of designing the optimal logical structure of the computing system [2, 3].

Let us assume that $\Phi = \{f_h; h = \overline{1, h}\}$ is a set of data processing problems in FNS and CS; hth problem from Φ is considered to be implementable, if the transformation of the set of values of X^h vector of input variables of the problem into the set of values of Y^h vector of output variables exists or can be synthesized:

$$Y^h = \pi^n(X^h);$$

where $X^h = (x_1^h, x_2^h, \ldots x_i^h, \ldots x_I^h)$—vector of input variables of the hth problem, and
$Y^h = (y_1^h, y_2^h, \ldots y_j^h, \ldots y_J^h)$—vector of output variables of the hth problem.

It should be noted that the distributed processing of information makes sense only if the transformation π^h does not satisfy a number of design specification requirements.

Domain D^h and range E^h of transformation π^h are defined as follows:

$$D^h = D_1^h * D_2^h * \cdots * D_i^h * \cdots * D_I^h, E^h = E_1^h * E_2^h * \cdots * E_j^h * \cdots * E_J^h,$$

where D_i^h—domain of variable x_i^h of vector x^h, and E_j^h—range of variable y_j^h of vector y^h, $x_i^h \in D_i^h$, $y_j^h \in E_j^h$, $i = \overline{1, I}, j = \overline{1, J}$.

Transformation π^h can be associated with multigraph $\Gamma = (A, \mathcal{A})$, which is an integrated structural graph and provides the vector of output variables y^h. Vertices $A = \{a_r; r = \overline{1, R}\}$ of the multigraph are data processing procedures, and edges are variables that are common to the corresponding procedures. It should be noted that any procedure $a_r \in A$ provides transformation $Y^r = a_r(X^r)$, for which

$$D_1^r * D_2^r * \cdots * D_i^r * \cdots * D_I^r \Rightarrow E_1^r * E_2^r * \cdots * E_j^r * \cdots * E_J^r,$$

where $X^r = (x_1^r, x_2^r, \ldots x_i^r, \ldots x_I^r)$—vector of input variables of procedures a_r,
$Y^r = (y_1^r, y_2^r, \ldots y_j^r, \ldots y_J^r)$—vector of output variables of procedures a_r,
D_i^r—domain of input variable x_i^r of procedures a_r, and
E_j^r—range of output variable y_j^r, procedures a_r.

$$x_i^r \in D_i^r, y_j^r \in E_j^r, i = \overline{1, I}, j = \overline{1, J}, r = \overline{1, R}.$$

Procedure a_r is a transformation of the set of input or intermediate variables of the problem into the set of intermediate or output variables of the problem. Set-based combination of procedures for the population of input and output variables defines set $Д$ of input, intermediate, and output variables of the problem, i.e.

$$Д^h = \bigcup_r \{X_r^h \bigcup Y_r^h\}.$$

Let us assume that $Q(A) = \{Q_\alpha; \alpha = \overline{1, (2^r - 1)}\}$—set of subsets A, where $Q_\alpha = \{a_r^\alpha\}$.

Let us assume that $Q_1(A)$ is some subset of set $Q(A)$, whose Q_α elements satisfy the condition

$$\bigcup_\alpha Q_\alpha = A, \quad \text{where } Q_\alpha \in Q_l(A). \tag{2.1}$$

In some cases, a constraint on the absence of duplication of elements (procedures) may be further imposed, i.e. $Q_\alpha \bigcap Q_{\alpha'} = 0, \alpha \neq \alpha'$.

Subset $Q_1(A)$ can be associated with aggregated graph $G = (\Gamma_\alpha, S)$, the vertices of which are subgraphs $\Gamma_\alpha = (Q_\alpha, D_\alpha), Q_\alpha \in Q_1(A), D_\alpha$—set of arcs incident to vertices $\{a_r^\alpha\}$; and arcs $S \in Д$—set of variables connecting the procedures of different Γ_α subgraphs between themselves.

Subgraphs Γ_α can be placed in matrix F with dimensionality $N * M$ for $\alpha = \overline{1, N * M}$.

$$F = \begin{vmatrix} \Gamma_{11} & \Gamma_{12} & \dots & \Gamma_{1m} & \dots & \Gamma_{MM} \\ \Gamma_{21} & \Gamma_{22} & \dots & \Gamma_{2m} & \dots & \Gamma_{2M} \\ \dots & \dots & \dots & & & \\ \Gamma_{N1} & \Gamma_{N2} & \dots & \Gamma_{Nm} & \dots & \Gamma_{NM} \end{vmatrix}$$

Let the elements of the matrix be subgraphs $\Gamma_{nm} = (Q_{nm}, D_{nm})$ forming graph $G = (\Gamma_{nm}, S)$, in which arc set S is as follows [4]:

$$S = \bigcup_{n=1}^{N-1} \bigcup_{m=1}^{M} \left\{ D_{nm} \bigcap \left[\bigcup_{m=1}^{M} D_{n+1,m} \right] \right\};$$

and the following condition is satisfied:

$$\bigcup_{m=1}^{M} \left\{ D_{nm} \bigcap_{m \neq m'} D_{nm'} \right\} = 0, \forall n, n = \overline{1, N} \tag{2.2}$$

where N—maximum number of stages of the integrated structural graph of the problem, and M—maximum number of parallel data processing branches.

Each subset $Q_1(A)$ and its corresponding graph G define a specific implementation of transformation π^h of the vector of input variables X^h into the vector of output variables Y^h.

Subgraphs $\Gamma_{nm} = (Q_{nm}, D_{nm})$ correspond to the transformations of the vectors of input variables X^{nm} into the vectors of output variables Y^{nm} of the Γ data processing graph. If transformation $Y^h = \pi^h(X^h)$ is representable as a graph

$$G = (\Gamma_{mm}, S) = ((Q_{mm}, D_{mm}), S),$$

then $\Gamma_{nm} = (Q_{nm}, D_{nm})$ are called the operating module (OM) of graph $\Gamma = (A, \varPi)$ of the problem corresponding to transformation π^h; the nth stage of the data processing graph is called set

$$\Im_n = \bigcup_m \Gamma_{nm}, \quad n = \overline{1, N}.$$

In this case, transformation π^h has the property of parallelism along M parallel branches. In accordance with the definition, the external interface of the integrated structural graph of the problem is $X^h \bigcup Y^h$.

Arc set S of graph G is called the interoperational information interface of the OM system of the data processing graph. Sets D_{nm} define the internal interface of OM [5].

Let us assume that $\varUpsilon$—the set of subsets $Q_1(A)$ satisfying condition (2.1). Then, in the general case, the problem of synthesizing the logical structure of the software and information support of the LAN-based CA FNS and CS can be written as follows:

$$\mathop{extr}_{Q_1(A) \subseteq Y_\varPi \subseteq Y} f[Q_1(A)],$$

where $f[Q_1(A)]$—function defined on set $\varUpsilon$, and $\varUpsilon_\varPi$—set of admissible partitions of graph Γ. Depending on the type and properties of the function, and the limitations that determine set $\varUpsilon_\varPi$, various problems of synthesizing the optimal logical structure of software and information support of the LAN-based FNS and CS can be set and solved.

The logical structure of software and information support of FNS and CS shown in Fig. 2.1 can be represented in the general case as a distributed data processing block diagram ("serial–parallel time grid").

A distinctive feature of FNS and CS relative to the ordinary ones is the information interconnection and interdependence of logical structures from the point of view of execution order. Non-receipt of information for at least one of the modules of the FNS and CS logical structure makes it impossible to solve the problem [2, 6].

Non-receipt of the input information array for the program module of the logical structure of the software and information support of the problem is possible for the following reasons: failure of the node in which the module, whose output information

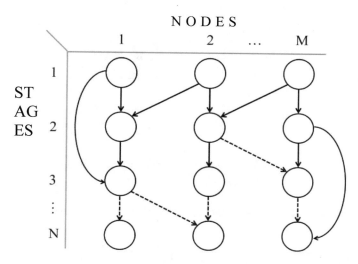

Fig. 2.1 Logical structure of software and information support of LAN-based FNS and CS

array is the input for the specified module, was executed; failure of the data transmission system connecting the nodes; destruction of the module, the output information array of which is the input for this module; and destruction of the information array.

Therefore, the redundancy of any part of the software module in other nodes of FNS and CS cannot lead to a noticeable increase in the problem-solving likelihood.

In this situation, the use of known redundancy models of a software module in data processing systems is hampered by the need to fulfill the requirements imposed on the parallelized problem-solving time. To overcome this difficulty, it is proposed to use structural and technological redundancy.

Taking into account the structural and technological redundancy of the software module, the composition of the logical structure is fully described by vector line $A = |A_1, A_2, A_3, ..., A_{m^*}, ..., A_M|$ consisting of Boolean matrices:

$$A_{m^*} = \begin{vmatrix} A_{11} & A_{12} & ... & A_{1M} \\ A_{21} & A_{22} & ... & A_{2M} \\ ... & ... & ... & \\ A_{N1} & A_{N2} & ... & A_{NM} \end{vmatrix}$$

where

$$A_{nm}^{m^*} = \begin{cases} 1, & \text{if module corresponding to subgraph } \Gamma_{nm} \\ & \text{is included in the composition of } \text{ПО and ИО } m^* \text{ node LAN,} \\ 0, & \text{otherwise}. \end{cases}$$

Let the set of LAN topologies be determined by a structured failure matrix:

$$\overline{B} = \begin{vmatrix} B_{11} & B_{12} & \ldots & B_{12}M \\ B_{21} & B_{22} & \ldots & A_{22}M \\ \ldots & \ldots & \ldots & \ldots \\ B_{M1} & B_{M2} & \ldots & B_{M2^M} \end{vmatrix}$$

where

$$B_{ij} = \begin{cases} 0, & \text{if } i \text{ node in } j \text{ combination is fault } y \\ 1, & \text{otherwise.} \end{cases}$$

Then the integrated structural–technological matrix is determined by the following formula:

$$H = A \times B = \left| H_1, H_2, H_3, \ldots, H_j, \ldots, H_{2M} \right|.$$

The structural and technological security of the software and information support of the problem being solved in the CA FNS and CS is determined by the following formula:

$$K = 100\% \sum_{j=1}^{2^M} y_j / 2^M$$

where

$$y_j = \begin{cases} 1, & \text{if } H_j\text{-unit vector,} \\ 0, & \text{otherwise.} \end{cases}$$

The structural and technological security determines the percentage of the number of network topologies that satisfy the user's requirements for solving the problem regarding the total number of LAN topologies.

The problem of structural and technological redundancy, in general, has the following form.

Find

$$\max_{A \in Q_{\text{Л}} \in Q} K,$$

where Q—set of matrices A that determine the composition of the software module of the logical structure taking into account duplication, and Q_D—set of admissible matrices A taking into account the limitations on the problem-solving time, the amount of external memory, etc.

Thus, the design of the optimal logical structure of software and information support of LAN-based CA FNS and CS is carried out on the basis of a structural approach that allows to formalize, algorithmize, and automatize the problem of designing the optimal logical structure of the CA FNS and CS. Since the redundancy of any part of the software module in other nodes of the FNS and CS cannot lead to an increase in the probability of solving the problem set, it is proposed to use a general mathematical model of structural and technological redundancy of the software and information support of the LAN-based CA FNS and CS [7].

2.1.2 Analysis of the Features of the Problem of Synthesizing Structural and Technological Redundancy and Its Representation as a Structural–Set Combinatorial Problem

The problem of synthesizing the structural and technological redundancy of the software and information support of the LAN-based CA FNS and CS refers to the combinatorial arrangement problems and has a number of features associated with constraints on the procedure arrangement variants. For example, constraint (2.2) prohibits the placement of information-related procedures at all stages due to the prohibition of the information transfer between LAN nodes within each of them. The same constraint prohibits placing, in a stage with a given sequence number, a procedure for which the input information elements are output information elements for procedures placed in stages with large sequence numbers. It should be noted that these features, which determine the technology of distributed data processing, greatly affect the system reliability indicators of quality.

Thus, taking into account the structural constraints in the problems of synthesizing the structural and technological redundancy of the software and information support of the LAN-based CA FNS and CS is reasonably necessary, and the effectiveness of the algorithm depends on how well it organically fits into the solution algorithm.

The proposed problems of synthesizing the optimal logical structure of FNS and CS are presented in the form of analytical models with Boolean variables. This form of representation is certainly not the only one. Many scientists emphasize the importance of combinatorial representation of certain classes of discrete optimization problems. In particular, it is noted that the set-theoretical approach contributes to the achievement of the greatest commonality of results; here are some combinatorial optimization problems: optimal error-correcting codes, block diagrams can be considered as extreme sets of subsets that satisfy certain set-theoretic constraints; this includes divisibility problems.

Combinatorial models can also be used to represent optimization problems that arise when optimally placed on graphs.

The analysis of the problems of synthesizing the structural and technological redundancy of the software and information support of the LAN-based CA FNS and CS shows the feasibility of studying them in a combinatorial representation taking into account the existing structural constraints.

The power of the combinatorial space of these problems has a significant impact on the efficiency of solving combinatorial problems. Therefore, a qualitative account of the natural features of the problems and, in particular, of structural constraints allows in many cases to significantly reduce the power of the combinatorial space and, therefore, improve the qualitative characteristics of the algorithm.

For the general statement of the problem of synthesizing the structural and technological redundancy of the software and information support of the LAN-based CA FNS and CS in the form of a structural set-based combinatorial problem, a number of definitions is to be introduced [8].

Definition 1 Let us assume that A—some discrete set. Set A is called structural if it is uniquely represented as a partition: $A = \{A_1, A_2, ..., A_n, ..., A_N\}$;

$$A_n \subset A, \forall n, n = \overline{1, N}; \; A = \bigcup_{n=1}^{N} A_n; \; A_n' \bigcap_{n' \neq n} A_n = 0, \forall n', n, n' = \overline{1, N}, n = \overline{1, N}.$$

Definition 2 Combinatorial space X is the collection of all combinatorial objects of a certain type from the elements of subset A_n, $A_n \subset A$; X—combinatorial space corresponding to set A.

In general, the following expression is true:

$$X \backslash X^* \neq 0, \text{ where } X^* = \bigcup_{n=1}^{N} X_n.$$

Set $A = \{a_1, a_2, ..., a_r, ..., a_R\}$ sets the ordered "seats" of elements $b_m, m = \overline{1, M}$ of set B. And $|A| \geq |B|$, i.e. $R \geq M$ (the number of procedures is not less than the number of network nodes); X—common space of arrangements with repetitions of M to R elements of set B, $|X| = M^R$. Obviously, for $M = 1$ space X coincides with set B, and for $M = R$ it covers the subspace of all permutations of elements of set B.

Then the problem of synthesizing the structural and technological redundancy of the software and information support of the LAN-based CA FNS and CS shows the feasibility of studying them in a combinatorial form as follows.

Find x^* from the equation:

$$f(x^*) = \min_{x \in X_{\mathcal{J}} \subset X} f(x),$$

where $X_{\mathcal{J}}$—region of feasibility set by the constraints of the problem, including structural constraints. Points $x, x \in X$ are set by tuples $x = (\alpha, ..., \alpha_r, ..., \alpha_R)$, where $\alpha_r = b_m$ with $\alpha_r \in A$, and $b_m \in B$; in this case, $\alpha_r = \alpha_m$ for $\alpha_r \in A$ is allowed.

The value of R plays the role of the parameter of set B when dividing the latter into subsets $B(R)$. Let also set N component values for parameter R, such that $\sum_{n=1}^{N} R_n = N$. The values of R_n correspond to subsets $A_n \subset A$, for which $R_n = |A_n|$ (in this case,

A_n is the set of procedures that make up the nth tier of the canonical structure of the software and information support of CA FNS and CS). Then set $B(R)$ can be divided into disjoint subsets $B_n(R_n) \subset B(R)$, for which $M_n = |B_n(R_n)|$, while it is advisable to divide the set of procedures of each tier into an equal number of components corresponding to the number of LAN nodes, i.e. $M_n = M$, $\forall n, n = \overline{1, N}$.

In this case, combinatorial space X is divided into a number of independent combinatorial subspaces.

Each subset $B_n(R_n)$ corresponds to some partial subspace X_n formed by the arrangement of elements and $b_n \in B_n$ in R_n positions, $|X_n| = M_n^{R_n}$.

Taking into account the structural constraints of the original problem, the objective functional, of which is defined on space X, allows replacing it with a similar problem of smaller dimension, the objective functional of which is defined on set $\widetilde{X}$ of selection arrangements of dimension N from a product of the form $X^1 \times X^2 \times \cdots \times X^n \times \cdots \times X^N$. Let us assume that $\tilde{x} = \left(x_1^1, \ldots, x_N^s\right) \in \widetilde{X}$—points of subspace $\widetilde{X} \subset X$, whose components are $x^n \in X^n$, $n = \overline{1, N}$. In this case, it is obvious that $\widetilde{X}$ coincides with the set of all permutations of X^n, $\left|\widetilde{X}\right| = N! \prod_{n=1}^{N} M_n^{R_n}$.

The gain from taking into account structural constraints can be determined by a value representing the power ratio of the combinatorial spaces of both problems:

$$(N \times M)^R / N! \prod_{n=1}^{N} M_n^{R_n} = N^R / N!$$

Thus, the analysis of the problem of synthesizing the structural and technological redundancy of the software and information support of the LAN-based CA FNS and CS showed the feasibility of studying it in a combinatorial representation taking into account the existing structural constraints. The power of the combinatorial space of these problems has a significant impact on the efficiency of solving combinatorial problems. Therefore, a qualitative account of the natural features of the problems and, in particular, of structural constraints allows in many cases to significantly reduce the power of the combinatorial space and, therefore, improve the qualitative characteristics of the algorithm.

2.1.3 Decomposition of the Problem of Synthesizing Structural and Technological Redundancy of Software and Information Support of Civil Aircraft LAN-Based Flight-Navigation and Computing Systems

The synthesis of the optimal structural and technological redundancy of the software and information support of the LAN-based CA FNS and CS can be carried out using precision and approximate methods. The exact solution to the problem can be obtained through its linearization.

However, the introduction of new (additional) problem variables leads to an increase in additional constraints and conditions related to their introduction, to an increase in the dimension of the original problem, which is not always justified and permissible from a practical point of view, since the use of standard computer software for solving linear programming problems may be ineffective and even unacceptable. That's why, for the problem of synthesizing the optimal structural and technological redundancy of the software and information support of the LAN-based CA FNS and CS, it is more feasible to use special methods for its solution reflecting the specifics of the problem.

The analysis of the set problem of synthesizing the optimal structural and technological redundancy shows that it is equivalent to the problem of dividing a graph into subgraphs while simultaneously taking into account the weights of vertices and arcs in the target function. Existing precise methods allow dividing a graph into no more than three subgraphs [9].

Methods for solving the problem of synthesizing the optimal structural and technological redundancy can be significantly simplified if there is at least one subproblem in the overall problem.

Therefore, it makes sense to distinguish between particular variants of the general problem in each specific case. For example, practice shows that there are possible variants for setting a general problem depending on the presence or absence of the possibility to choose the number of LAN nodes, as well as using a given logical structure of the software and information support of the LAN-based CA FNS and CS.

The proposed combinatorial form of the problem of synthesizing the optimal structural and technological redundancy makes it possible to most fully take into account its specifics due to structural constraints and to develop decomposition-based algorithms.

The parallel decomposition algorithm is based on the fact that for all points $\tilde{x} \in \tilde{X}$, functional $f(x)$ can be represented as

$$f(\tilde{x}) = \sum_{n}^{N} f_n(x_n) + \Delta f(\tilde{x}), \tag{2.3}$$

where $f_n(x_n)$—functional defined on partial subspace X_n, and $\Delta f(\tilde{x})$ is defined for all $\tilde{x} \in \tilde{X}$. In this case, the original problem with target functional $f(n)$ is replaced by a similar one with functional $f(\tilde{x})$.

The parallel decomposition algorithm of the problem with the constraints introduced can be constructed in two forms: as single-stage, according to which functional $f(x)$ is minimized on the whole subspace $\tilde{X}$, and as two-stage, according to which $f(x)$ is minimized in parts, respectively, on the right side of expression (2.3).

The second type of algorithm is faster. It allows parallelizing the computational process and is reduced to solving a number of auxiliary and one main decomposition problems. Auxiliary combinatorial problems have the following form:

$$f_n(x_n^*) = \min_{x_n \in X_Д \cap X_n} f_n(x_n).$$

The solution to these problems for $n = \overline{1, N}$ is the following: $x_1^*, \ldots, x_n^*, \ldots, x_N^*$. The main decomposition problem is constructed to minimize functional $\Delta f\ (\tilde{x})$, the value of which depends on the sequence of components $x_n^j \in X_j$. However, it is solved only in part $\widetilde{X}^0$ of space X, $\widetilde{X}^0 \subset X$ consisting of points $x = (x_{1L}^*, \ldots, x_{Ns}^*), x_{1L}^* \in X_L, x_{Ns}^* \in x_s$, and its combinatorial form is as follows:

$$\Delta f(\overset{*}{\tilde{x}}) = \min_{\tilde{x}_n \in X_Д \cap X^0} \Delta f(\tilde{x}).$$

Along with the above-noted positive properties of the combinatorial form of representing the problem of synthesizing the optimal structural and technological redundancy, we should note its negative property of masking specific features of descriptive components expressing their physical meaning. Therefore, the proposed decomposition models for synthesizing the optimal structural and technological redundancy shall be represented in the form of discrete optimization models.

At the first stage, N problems for synthesizing the optimal logical structure of the software and information support of CA FNS and CS are solved for each tier according to the criterion of uniform complexity. The result of solving these problems is the procedural and informational structure of the operational modules, as well as their total number. Then, the problem of synthesizing the optimal logical structure of the software and information support of the LAN-based CA FNS and CS is solved taking into account duplication of software modules. Copies of the software module and the information array arranged in LAN nodes will be a structural and technological redundancy [10].

In the case when the possible topologies of a homogeneous LAN are not equally probable, the synthesis of the structural and technological redundancy shall be performed taking into account the probabilities of a failure of the LAN nodes and the probability of a failure of communication channels between the LAN nodes. Then, it is advisable to accept the minimum probability of a failure to solve the user's problem as a criterion for the structural and technological redundancy optimization problem.

Formulation of the problem requires determining the following values:

P_m^*—probability of failure of the m^*th LAN node;

P_{m1m2}—probability of failure of the information channel connecting the nodes m_1, m_2;

$V = (V_m^*)$—vector of the amount of external memory of the LAN nodes, where V_m^*—amount of available external memory of the m^*th LAN node;

$F = (f_{nm})$—vector of volumes of the CA FNS and CS logical structure modules, where f_{nm}—volume of the nmth module;

$\Phi = \left(\varphi_{nm}^{n'm'}\right)$—vector of volumes of the information array, where $\varphi_{nm}^{n'm'}$—volume of the information array common to the modules nm and $n'm'$;

$C = \left(C_{m_1^* m_2^*}\right)$—vector of the bandwidth of the communication channels between LAN nodes m_1^*, m_2^*, where $C_{m_1^* m_2^*}$—bandwidth of the communication channel between nodes m_1^*, m_2^*;

$\Lambda = (\lambda_m^*)$—vector of the performance of LAN nodes, where λ_m^*—performance of the m^*th LAN node;

$Q = (Q_{nm})$—vector of the complexity of the modules of the logical structure of software and information support, where Q_{nm}—complexity of the nmth module;

$$P_{m_1^* m_2^*}^{nmn'm'} = \begin{cases} 1, if\ \varphi_{nm}^{n'm'} = 0, \\ P_{m_1^* m_2^*}, if\ \varphi_{nm}^{n'm'} \neq 0\ u\ X_{nm}^{m_1^*} + X_{nm}^{m_2^*} = 1 \end{cases}.$$

Values $P_{m_1^* m_2^*}^{nmn'm'}$ establish the relationship between the information array transmitted from module nm to module $n'm'$ and the probability of failure of the information channel connecting nodes m_1^*, m_2^*, through which this information array is transmitted.

The problem variables are [5, 11]

$$X_{nm}^{m^*} = \begin{cases} 1, & \text{if } nm \text{ module is placed at } m^* \text{ node } BC, \\ 0, & \text{otherwise} \end{cases}$$

$$\varphi_{nm}^{m^*} = \begin{cases} P_{m^*}, if\ X_{nm}^{m^*} = 1, \\ 1, if\ X_{nm}^{m^*} = 0 \end{cases}.$$

The mathematical model of optimization of the structural–technological redundancy of the software and information support of the CA FNS and CS in a homogeneous LAN according to the criterion of the minimum probability of failure to solve the problem has the following form.

Find [12]

$$\min_{\{X_{nm}^{m^*}\}} \prod_{n=1}^{N} \prod_{m=1}^{M} \left[\prod_{m^*}^{M} \varphi_{nm}^{m^*} \prod_{n'=1}^{N} \prod_{m'=1}^{M} \prod_{m_1^*=1}^{M^*} \prod_{m_2^*=1}^{M^*} P_{m_1^* m_2^*}^{nmn'm'} \right],$$

under the following constraints:

- on structural duplication of modules $X_{nm}^{m_1^*} X_{n'm'}^{m_2^*} = 0$ for $\forall n, m, n', m', m_1^*, m_2^*$, for which conditions $C_{m_1^* m_2^*} = 0$, $\varphi_{nm}^{n'm'} \neq 0$ are met;
- on the distribution of individual modules for individual nodes $X_{nm}^{m^*} = 1$, for the selected nmth operational modules and m^*th nodes;

– on the longest possible time to solve the problem (2.4).

$$\sum_{n=1}^{N} \max_{\{m^*\}\atop\{m\}}\left[X_{nm}^{m^*}\theta_{nm}\lambda_{nm}\right] + \sum_{n=1}^{N}\sum_{n/}\sum_{m=1}^{M}\sum_{m/=1}^{M}\max_{\{m_1^*\}\atop\{m_2^*\}}\left[X_{nm}^{m_1^*}X_{n/m/}^{m_2^*}\varphi_{nm}^{n/m/}\frac{1}{C_{m_1^*m_2^*}}\right] \leq T^*,$$

where T^*—maximum possible time to solve the problem;

– on the maximum amount of external memory of LAN nodes

$$\sum_{n=1}^{N}\sum_{m=1}^{M} x_{mn}^{m} f_{mm} \leq V_{m}*, \forall m^*, m^* = \overline{1, M}. \tag{2.4}$$

The problem set is a non-linear programming problem with Boolean variables. It can be reduced to an integer optimization linear programming problem by logarithming the target function and simplifying constraint (2.4) to a linear one.

That's why the problem of synthesizing the optimal structural and technological redundancy of the software and information support of the LAN-based CA FNS and CS can be solved through its linearization. The analysis of the problem set shows that it is equivalent to the problem of dividing a graph into subgraphs while simultaneously taking into account the weights of vertices and arcs in the target function. The proposed combinatorial form of the problem under consideration made it possible to most fully take into account its specifics due to structural constraints and to develop decomposition-based algorithms. The developed decomposition models for synthesizing the optimal structural and technological redundancy are represented in the form of discrete optimization models.

Due to the fact that possible topologies of a homogeneous LAN may not be equally probable, a mathematical model has been developed for optimizing the structural and technological redundancy of the FNS and CS software and information support according to the criterion of the minimum probability of failure to solve the problem.

Thus, the use of structural and technological redundancy of the software and information support of the problems being solved in FNS and CS allows increasing the security of the logical structures of CA FNS and CS under the effect of destabilizing factors taking into account the constraints on the maximum time to solve problems. The implementation of the proposed model for synthesizing the structural and technological redundancy allows, for specific LANs, determining the redundancy rates for the structure and amount of redundancy of the software and information support of the distributed problems of LAN-based CA FNS and CS taking into account the established lower security limit. In the case of parallel processing, such rates can be also defined for the reserve amount of external memory.

2.2 Development of Virtual Recovery Data Redundancy Models of Civil Aircraft LAN-Based Flight-Navigation and Computer Systems

LAN-based CA FNS and CS are sometimes used as organizational and technical forms of managing real-time processes. Such systems demand a higher speed of data processing under real-life conditions. For example, quick-response FNS and CS built to support decision-making in emergency situations demand both higher integrity of the data used in these systems and higher speed of their processing. As a rule, these systems are used as an organizational–technical form of the interconnection of geographically distributed controls of various kinds of activity within one or several duty dispatcher services.

The solution to the problem of data integrity in such systems is relevant and requires consideration of a number of factors, such as distributed data processing, severe time constraints for receiving a query response when making a decision, and large amounts of data for visualizing the learning environment.

The output information refers to information obtained as a result of the FNS and CS functions and issued to the object of its activity, users, or other systems. The quality of the output information in the FNS and CS is a set of information features that determine its suitability to meet the users' requirements when solving the problems facing them.

One of the main system methods for improving the quality of the output information in the FNS and CS aimed at improving the probability–time characteristics of the systems' functioning is backup and recovery of data.

When solving problems of improving the quality of output information using backup and recovery methods, functioning properties of the LAN-based FNS and CS allow, in addition to traditional redundancy introduction means, introducing and effectively using redundancy, which can be defined as a virtual recovery one. It takes into account the property of information reliability not only as a property of accuracy (i.e. compliance of data received by the consumer with data generated at the source), but also as a property of relevance (maintaining a sufficient degree of compliance with the real state of accounting objects at the time of using this information).

The essence of the virtual recovery backup is that in most cases the problems solved by geographically distributed controls allow pre-allocating the necessary data sets (intermediate arrays) and distributing them to the aircraft nodes for immediate use in the future. In contrast to the structural and technological backup, the virtual recovery backup can include both the data itself and its copies and/or pre-history.

Thus, in some cases the following is allowed:

$$|X(t) - X(t - \tau)| \leq \delta, \delta > 0, \forall \tau, \tau \in [0, t], \tag{2.5}$$

where $X(t), X(t - \tau)$—the corresponding values of any parameter of the information element at time t and $(t - \tau)$, and δ—absolute threshold of the permissible limits of deviations of the used information from the real one. The fulfillment of inequality

(2.5) determines the relevance of the information element at interval $T = [0, t]$. In this case, the degree of backup virtuality is the probability of irrelevance of all its constituent elements at interval T.

One of the main problems of virtual recovery data backup solved at the pre-project analysis stage is the problem of determining the optimal content of the virtual recovery backup and its arrangement on the LAN nodes.

Let us assume that M—number of LAN nodes, N—number of information elements of the system, α_n—relative frequency of adjustments of the nth information element (number of adjustments for a given time period T), $t^n_{mm'}$—time of transmission of the nth information element from the mth to the m'th LAN node, b_n—volume of the nth information element, $W = \|\lambda_{nm}\|$—matrix of interconnections of information sources (LAN nodes) and information elements,

$$\lambda_{nm} = \begin{cases} 1, \text{ if the source of the } n\text{-th information element} \\ \quad \text{is the } m\text{-th network node;} \\ 0, \text{ otherwise} \end{cases}$$

and $C = \|c_{nm}\|$—gain matrix, where c_{nm}—weighted estimates of the gain obtained by the user as a result of placing the nth information element in the mth LAN node network, $c_{nm} \in [0.1]$, $\forall n, n = \overline{1, N}$, $\forall m, m = \overline{1, M}$. Methods for constructing matrix C are determined by the specific conditions of the use of FNS and CS. In particular, an estimate of the gain obtained by the user as a result of placing the nth information element in the mth LAN node can be made up of objective estimates (for example, time of transfer of the nth information element to the mth node from other LAN nodes) and the subjective estimate of negativism in relation to the need to transfer the nth information element to the mth LAN node.

Then, using the following variables [13]

$$x_{nm} = \begin{cases} 1, \text{ if the } n\text{-th information element is placed} \\ \quad \text{in the } m\text{-th LAN node;} \\ 0, \text{ otherwise,} \end{cases}$$

the degree of backup virtuality $P(x_{nm})$ can be determined by the following formula:

$$P(x_{nm}) = 1 - \prod_{n-1}^{N} \left(1 - a_n \sum_{m=1}^{M} \sum_{m'}^{M} x_{nm} \lambda_{nm'} t^n_{mm'} \right).$$

In cases when geographically distributed systems consist of elements that are homogeneous in terms of the degree of risk of emergency situations, it is advisable to use the maximum criterion of uniform gain distribution over the LAN nodes, the criterion of the minimum degree of backup virtuality, etc. as the virtual recovery backup synthesis criteria.

The problem of designing a virtual recovery backup according to the first criterion is as follows.

Find

$$\max \quad \min_m \sum_{n=1}^{N} c_{nm} x_{nm}, \tag{2.6}$$

under the following constraints:

– on the degree of backup virtuality

$$1 - \prod_{n=1}^{N} \left(1 - \alpha_n \sum_{m=1}^{M} \sum_{m=1}^{M} x_{nm} \lambda_{nm'} t_{mm'}^{n} \right) \leq P^*, \tag{2.7}$$

where P^*—maximum allowable degree of backup virtuality;

– on the relative time of adjustments of information elements

$$\sum_{n=1}^{N} \sum_{m=1}^{M} \sum_{m'=1}^{M} x_{nm} \alpha_n \lambda_{nm} t_{mm'}^{n} \leq T_{cor}^*, \tag{2.8}$$

where T_{cor}^*—maximum allowable relative time of adjustment of the information element (at interval T);

– on the amount of external memory of the mth LAN node

$$\sum_{n=1}^{N} x_{nm} b_n \leq B_m^*, \forall m, m = \overline{1, M}, \tag{2.9}$$

where B_m^*—maximum allowable amount of memory of the mth node for storing information;

– on the lack of duplication of the information element in the LAN nodes

$$\sum_{n=1}^{N} x_{nm} = 1, \forall m, m = \overline{1, M}. \tag{2.10}$$

Statement 1

The solution to problems (2.6)–(2.10) is admissible, if $P^* \Rightarrow 1$ and constraints (2.8)–(2.10) are met.

Evidence

Let us assume that $\varphi_n = \sum\limits_{m=1}^{M} \sum\limits_{m'=1}^{M} x_{nm} \alpha_n \lambda_{mm'} t_{mm'}^n$. Then constraints (2.7), (2.8) will have the following form:

$$\begin{cases} 1 - \prod\limits_{n=1}^{N} (1 - \varphi_n) \le P^* \\ \sum\limits_{n=1}^{N} \varphi_n \le T_{\kappa op}^* \end{cases} \tag{2.11}$$

After taking the logarithm of the first system inequality (2.11) and expanding the logarithm function in a power series, we will consistently get

$$\begin{cases} \sum\limits_{n=1}^{N} \ln(1 - \varphi_n) \le \ln(1 - P^*); \\ \sum\limits_{n=1}^{N} \varphi_n \le T_{\kappa op}^*; \end{cases} \qquad \begin{cases} \sum\limits_{n=1}^{N} \varphi_n \le -\ln(1 - P^*) - \varepsilon, \\ \sum\limits_{n=1}^{N} \varphi_n \le T_{\kappa op}^*, \end{cases}$$

where ε—residual member of the series.

Obviously, constraint (2.8) is more stringent with $P^* \Rightarrow 1$. The proposed constraint rigidity analysis can be used to reduce the dimensionality of problems with specific input data.

Modern LAN-based FNS and CS of an organizational type are critical to the amount of information transmitted via communication channels. Therefore, the most urgent problem can be the problem of finding

$$\max_{x_{nm}} \min_{m} \sum_{n=1}^{N} c_{nm} x_{nm}$$

with constraint (2.8). This problem is reduced to the maximization problem by introducing additional variables $y_m = \{0.1\}$. It has the following form.

Find

$$\max_{\{x_{mm} y_m\}} \sum_{m=1}^{M} y_m \sum_{n=1}^{N} c_{nm} x_{nm}$$

under the following constraints:

$$-\sum_{m=1}^{M} y_m \sum_{n=1}^{N} c_{nm} x_{nm} \le \sum_{n=1}^{N} c_{nm} x_{nm}, \forall m, m = \overline{1, M}$$

$$-\sum_{m=1}^{M} y_m = 1$$

$$-\sum_{n=1}^{N}\sum_{m=1}^{M}\sum_{m'=1}^{M} x_{nm}\alpha_n\lambda_{nm'}t_{mm'}^{n} \leq T_{\kappa op}^{*}.$$

Let us assume that $t_{mm'}^{n} = \text{const}$, $\forall m$, $m = \overline{1, M}$, $\forall m'$, $m' = \overline{1, M}$, $m' \neq m^*$, $\forall n$, $n = \overline{1, N}$. Then the solution to the problem of determining the optimal content of the virtual recovery backup and its placement in the aircraft is reduced to solving M multi-dimensional knapsack problems, which are formulated as follows.

Find

$$\max \sum_{n=1}^{N} c_{nm^*}x_{nm^*}, \forall m^*, m^* = \overline{1, M}$$

under the following constraints

$$-\sum_{m=1}^{M}\sum_{n=1}^{n}\alpha_n x_{nm} \leq D; D = T_{\kappa op}^{*}/t_{mm'}^{n}$$

$$-\sum_{n=1}^{N}(c_{mm^*}+\alpha_n) + \sum_{n=1}^{N}(\alpha_n - c_{nm'})x_{nm'}$$

$$+ \sum_{\substack{m=1 \\ m \neq m' \\ m \neq m^*}}^{M}\sum_{n=1}^{N}\alpha_n x_{nm} \leq D, \forall m', m' = \overline{1, M}, m' \neq m^*.$$

The result of solving the problem of synthesizing virtual recovery backup is the optimal information set of data arrays arranged on LAN nodes according to specified criteria. The use of virtual recovery data backup in the LAN-based FNS and CS, based on a more flexible use of the concept of information quality, improves the speed of data processing, as well as its integrity under the effect of destabilizing factors.

Thus, the solution to the problem of data integrity in the CA FNS and CS is based on the backup and recovery of data. When solving the problem of information integrity using backup and recovery methods, it is proposed to use a virtual recovery backup, which includes both the data itself, as well as its copies and/or pre-history. As a result, the problem of determining the optimal content of the virtual recovery backup and its arrangement on the LAN nodes is solved at the stage of pre-project analysis. In cases when FNS and CS consist of elements that are homogeneous in terms of the degree of risk of emergency situations, it is proposed to use the maximum criterion of uniform gain distribution over the LAN nodes, the criterion of the minimum degree of backup virtuality, etc. as the virtual recovery backup synthesis criteria [14].

2.3 Methods for Decomposing Civil Aircraft Flight-Navigation and Computing Systems

The problem of network synthesis is simplified if the specifics of the conditions for its creation and operation are taken into account. The specificity is that the organizational structure of the network includes a number of objects located at a considerable distance from each other. The specified objects, in accordance with the set information needs, are equipped with communication (data transmission) facilities and CS of the required performance, thus naturally forming network nodes. These circumstances suggest that the overall structure of the network is as set.

The structure and parameters of the flow intensity in computer networks largely depend on the organizational solutions that determine the distribution of software and data in the network taking into account the provision of the necessary degree of system survivability, as well as the corresponding informational interchange during its operation. Such organizational solutions can be taken on the basis of the analysis of the corresponding optimization models only. Due to the multidimensionality of the information needs of FNS and CS objects and the complexity of information processing procedures, the optimization problems arising in this case have a higher dimension, and their solution is associated with overcoming significant computational difficulties.

One of the essential factors determining the stability of the computational flow intensity and the functioning of the system as a whole against the action of destabilizing factors is the rational arrangement of information resources of the network (software, information array, and their recovery backup). The solution to this problem shall contribute to the selection of optimal engineering solutions at various stages of design, operation, improvement, and development of networks. Their solution shall be ensured by both analysis and optimal synthesis of the system of computational tools and their components. In addition to that, there are complex problems of optimal system synthesis. In particular, this refers to the network-level synthesis of the system. Such a synthesis, taking into account the choice of the topology of the data transmission network, leads to a model of such a high dimensionality that it presents considerable difficulties with the existing level of development of mathematical methods and computer technology. Consequently, sufficiently compact statements of the corresponding optimization problems can be obtained only with the introduction of some simplifying assumptions. These assumptions are determined by the specifics of the organizational support of the FNS and CS, where the main solutions on the topology and means of a certain subsystem are taken based on the needs of the higher level, and are mandatory for the FNS and CS. In addition, it is necessary to take into account that the structure of network nodes, in general, is a homomorphic image of the structure of the corresponding objects of the CA FNS and CS. In this situation, the main problems of static optimization of computing flow intensity at certain levels of their representation are greatly simplified and can be formulated as mathematical programming problems.

To reduce the dimensionality of the computational flow intensity optimization problems, it is proposed to consider the FNS and CS as a set of nested control loops. The main purpose of this partition is such an organization of the system which leads to the need to make changes either in one of its elements, or, in extreme cases, in the minimum number of them [4, 7, 15].

The system shall be partitioned in such a way as to provide the territorial distribution of the aircraft ensuring the necessary survivability by expanding its reconfiguration capabilities and increasing its reliability. This kind of partition can be called vertical. With vertical partition, the data flow cross-section is recommended to be performed at those points where the amounts of transmitted or stored data are minimal. These points are often associated with major events or solutions about the ways of transmission.

In many applications, a number of data processing functions can be implemented in parallel without mutual time synchronization. In such cases, horizontal partition can be used. It is advisable with full autonomy of processing steps, i.e. if they do not have a data link or precedence-following relationship in managing data flows. Horizontal partitioning is also possible in the case of parallel event processing when the processing of any of them is not correlated with the processing of others.

Each control loop shall correspond to its own detailed description of the computational flow intensity, which increases as it moves downward (in the direction of a more detailed description of the processes), which allows the harmonization of the main elements of the computational flow intensity at an appropriate level. The number of control loops shall be determined based on the practical needs of the study being conducted. With this approach, the solution to any rather complex problem can be achieved as a result of the consistent refinement of the values of the system parameters and its structural components using calculations through a set of mathematical models.

Representation of the control system as a set of nested loops greatly facilitates the organization of network design and reduces the duration and cost of development. It becomes possible to rationally distribute the efforts of developers to solve private problems depending on their importance. The number of components at different levels, taking into account the complexity of relationships, can be analyzed to estimate the amount of the work done and predict its prospects in terms of time and complexity, as a result of which the reliability of the state control and design process is improved.

The proposed approach allows designing complex information computing flows on a "top-down" basis from the perspective of purpose and the best solution to the main target problem of the entire system. This ensures the conceptual unity of the information computing flow and the possibility of rational allocation of resources as the system is decomposed. Although the division into levels requires some costs, in general, resources are used more efficiently due to the economical arrangement of the information computing flow at each level. At the same time, possible partitioning criteria may include the minimum cost of data processing and transmission, the total time of solving problems, network bandwidth, failure probability, and other criteria.

At the design stage, it is proposed to solve the problems of optimizing the intensity of the computational flow (distribution of software modules and information arrays, as well as their backup) according to the "top-down" principle. For the network level, they are generally defined as problems of the optimal distribution of the storage and processing functions within the system taking into account the appropriate choice of the type of computing tools in the network nodes. At the same time, the cost of network resources for the implementation of a given set of the intensity of the computational flow is selected as the target function of the problem. For a given level, these costs can be associated with the average amount of information transmitted in the network when all the computational flow intensities from the set are met. The latter is explained by the fact that meeting a certain computational flow intensity is associated with a quite definite scope of computation, so that any gain in optimizing the intensity of the computational thread can be obtained only by reducing the amount of information transmitted within the network. And this, in turn, becomes possible with such distribution of information in the system when frequently used programs and data are stored wherever possible, where the need for them arises. The solution of this task allows reducing the time of data transmission and recovery of the destroyed information, and, consequently, the time for the implementation of the applied functions of CA FNS and CS. In addition, estimating the amount of transmissions with the optimal arrangement option allows making an economical choice of data channels of a certain bandwidth at the FNS and CS building stage and reduce the damage from the implementation of attacks on aircraft based on the analysis of the activity of network nodes [16].

In the lower level control loops, the problems of static optimization of the computational flow intensity are based on the determination of composition and structure of the CS and the distribution of problems (software), information array (databases), and their recovery backup between several complexes taking into account their priority and intensity of the solution, constraints on the amount of memory, and time to solve each problem, as well as on the determination of the volume of recovery backup of each software module and information required to ensure the specified level of the information integrity indicator. Moreover, the problem of optimization of the computing resource at this level shall be solved depending on the operating conditions of the FNS and CS, according to the criteria of maximum probability of solving all problems and the minimum time to solve them, which will make it possible to select the media required for storing software modules, information arrays, and their backups at the network design stage.

Thus, the solution to the sequence of optimization problems allows determining and specifying the arrangement of information resources at the CA FNS and CS design stage.

At the stage of operation, the problem of distributing software, information arrays, and their computing resources is solved when individual components of the system fail and when new application software is commissioned. To increase the stability of the computational flow intensity at the stage of network operation, it is advisable to distribute (redistribute) information resources according to the "bottom-up" principle.

The main problem in optimizing the intensity of the computational flow is the organization of problem distribution on efficient computers. The solution to this problem is associated with the characteristics of the problems and the requirements for the type of degradation. In case of a failure of individual computers, problems solved by the system may be redistributed among operable computers. This allows maintaining the operability of the system by reducing any operational quality indicators within acceptable limits. Systems in which this feature is realized are called gradual degradation systems. There are two types of degradation that are possible during the transition of the system from one state to another and the redistribution of problems within it:

- performance degradation, i.e. decrease in system performance due to a decrease in the number of parallel-operating computers. When redistributing problems in the system, the number of problems solved successively by each of the operable computers may increase due to the fact that it takes on the problems of the failed computers. Hence, an increase in the time for processing problem-solving application is possible, and for a control system—increase in the time of response to a change in input actions of the system;
- algorithmic (functional) degradation, which can be characterized by the following:

  - decrease in the number of problems solved by the system or decrease in their total weight, where the weight of each problem refers to a value characterizing the importance or criticality of each problem in terms of management processes performed;
  - decrease in the complexity of the algorithms performed by reducing their "quality".

The following problem distribution method is proposed. Initially, the problems solved by the computing system are distributed according to the criterion of uniform computer loading. With a decrease in computing resources due to failures of individual computers, when the distribution by the criterion of uniform loading is impossible, the distribution is carried out by the criterion of the maximum importance of problems.

In case of a failure of the significant number of computers of the control loop, i.e. a significant reduction in computing power, the distribution (redistribution) of software modules, information arrays, and their recovery backup on operable computers is performed depending on the need and feasibility of problem-solving, i.e. situational management of computing resources is carried out. If it is impossible to solve the problem in this control loop, a transition to the adjacent "horizontal" or "vertical" level occurs.

The distribution of software, information arrays, and their backup between the highest level loops is carried out taking into account the minimum amount of transmitted information. In this case, the correct and timely solution to these problems contributes to maintaining the operability of the system with the same performance and throughput.

The proposed approach and methods for the distribution of information resources, definitely, do not cover all possible situations related to static and dynamic optimization of the computational flow intensity at various levels of their presentation. These problems demonstrate a general idea, which consists in a preliminary determination of such a computational flow intensity organization at the system design and setup stage, which would allow in a certain sense implementing the computational flow intensities set in the system in an optimal way.

Thus, one of the essential factors determining the stability of the CFI and the functioning of the system as a whole against the effects of destabilizing factors is the rational allocation of information resources of the computer network. To reduce the dimensionality of the CFI optimization problems, it is proposed to consider the MAN as a set of nested control loops. With this approach, the solution can be achieved as a result of the consistent refinement of the values of the system parameters and its structural components using calculations through a set of mathematical models.

2.4 Mathematical Model for Optimizing Recovery Information Backup in Modern Civil Aircraft Flight-Navigation and Computing Systems

2.4.1 General Mathematical Model for Optimizing Recovery Information Backup in Modern Flight-Navigation and Computing Systems

Let a network consisting of L elements, each of which has m_j $(j = 1, …, L)$ information processing points, be given.

The network solves K problems that use data from M information arrays. At each hth point, jth element $(j = 1, 2, …, L)$ $(h = 1, 2, …, m_j)$, a strictly defined range of problems is solved using certain information arrays and generating the corresponding requests (messages).

The distribution of software modules and information arrays over network nodes is determined by the distribution plan defined by the matrices

$$X = \left\| x_{kj} \right\|, Y = \left\| y_{fj} \right\|, \Psi = \left\| \psi_{kj} \right\|, \Phi = \left\| \varphi_{fj} \right\|,$$

where

$$x_{kj} = \begin{cases} 1, & \text{if } k \text{ program module is placed on } j \text{ element,} \\ 0, & \text{otherwise} \end{cases}$$

$$y_{fj} = \begin{cases} 1, & \text{if } f \text{ information array is located on } j \text{ element,} \\ 0, & \text{otherwise} \end{cases}$$

$$\psi_{kj} = \begin{cases} 1, & \text{if reserve of } k \text{ program module is placed on } j \text{ element,} \\ 0, & \text{otherwise} \end{cases}$$

$$\varphi_{fj} = \begin{cases} 1, & \text{if reserve of } f \text{ information array is placed on } j \text{ element,} \\ 0, & \text{otherwise} \end{cases}$$

$$k = 1, 2, \ldots, K, \, f = 1, 2, \ldots, M, \, j = 1, 2, \ldots, L.$$

Let's define z_k, $\bar{z}_f$—volume of the recovery backup of the kth software module and the fth information array (number of copies (pre-histories) of the kth software module, the fth information array) ($k = 1, 2, \ldots, K; f = 1, 2, \ldots, M$), respectively.

The destruction of a software module or an information array means its state in which the module or array can no longer be used by the system to obtain the required output as a result of the loss of part or all of the information contained in it.

In case of destruction of the software module (information array), it is restored using its first copy (pre-history). If a copy is destroyed, it is restored using the next copy (pre-history), etc.

A request (message) for solving the kth task ($k = 1, 2, \ldots, K$) may not be processed due to the following:

- destruction of the kth software module or the lth information array used in its solution, and the impossibility of its restoration due to the destruction of the recovery backup during storage or use;
- unsatisfactory state of the data transmission channels during the transmission of the request or the restoration of corrupted information.

Then, according to the above scheme, the expressions for determining the probability of a successful solution P_{jhk}, the time to solve problems, and the amount of information flow circulating in the network when each problem is being solved by each subscriber

$$\Lambda_{jhk}(j = 1, 2, \ldots, L; k = 1, 2, \ldots, K; h = 1, 2, \ldots, m_j)$$

have the following forms:

$$P_{jhk} = \tau_{jhk} \sum_{l=1}^{L} P_{jhkl}^P P_{jlhk}^n x_{kl} \prod_{f=1}^{M} \sum_{r=1}^{L} y_{fr} \left(P_{lkfr}^0 \overline{P}_{lrkf}^n \right)^{q_{jhkf}}, \tag{2.12}$$

$$T_{ihk}^{\text{рещ}} = \frac{1}{\tau_{jhk}} \sum_{l=1}^{L} x_{kl} \left(T_{jlhk} + t_{jhkl}^{\text{рещ}} + Q_{jhkl} \, t_{kl}^{\text{в}} + \sum_{f=1}^{M} q_{jhkf} \sum_{r=1}^{L} (\overline{T}_{lrkf} + \overline{Q}_{lkfr} \, \bar{t}_{fr}^{\text{в}}) y_{fr} \right), \tag{2.13}$$

$$\Lambda_{jhk} = \lambda_{jhk} \left(\sum_{i=1}^{L} \left\{ \begin{array}{l} x_{ki} \left[F_{ji} 1_{jhk}^3 + F_{ij} l_{jhk}^c + Q_{jhki} \sum_{l=1}^{L} \psi_{kl} \left(F_{il} 1 k^B + F_{li} u_k \right) \right] + \\ + \sum_{f=1}^{M} \sum_{r=1g=1}^{Lq_{jhhf}} y_{fr} \left[F_{ri} \bar{1}_{kgf}^3 + F_{ir} \bar{1}_{kgf}^c + \overline{Q}_{ikfr} \sum_{l=1}^{L} \varphi_{fl} \left(F_{rl} \bar{l}_f^B + F_{1r} \delta_k \right) \right] \end{array} \right\} \right) \tag{2.14}$$

where $P_{jihk}^{\Pi}(\overline{P}_{lrkf}^{\Pi})$—probability of successful transmission of information between nodes $j(l)$ and $i(r)$ in the solution to the kth problem (reference to the fth information array) by the hth subscriber (reference from the kth software module) of the jth element (located in the lth node).

$$P_{jihk}^{\Pi} = P_{jihk}^{3} P_{ijhk}^{c},$$
$$\overline{P}_{irkf}^{\Pi} = \overline{P}_{lrkf}^{3} \overline{P}_{rlkf}^{c};$$

$P_{ijhk}^{3}(\overline{P}_{lrkf}^{3})$, $P_{ijhk}^{c}(\overline{P}_{rlkf}^{c})$—probability of communicating a request for the solution of (access to/reference to) and the message containing solution (reference) results of the kth problem (reference to the fth information array) by the hth subscriber (reference from the kth software module) of the jth element (located in the lth node) in the ith network node (located in the rth node), respectively;

P_{jhkl}^{P}, P_{lkfr}^{0}—probability that the kth software module stored on the lth element is not run when referenced by the hth subscriber, the jth element is not destroyed or will be successfully restored, and probability that the fth information array stored on the rth element is not run when referenced by the kth software module located at the lth element, not destroyed or will be successfully restored, respectively:

$$P_{jhkl}^{P} = 1 - Q_{jhkl}\left[1 - \sum_{g=1}^{L} \psi_{kg}\left(1 - Q_{kg}^{P}\right)P_{\lg k}^{\Pi B}\left(1 - \rho_{kl}^{k}\right)\right], \qquad (2.15)$$

$$P_{jhkl}^{o} = 1 - \overline{Q}_{lkfr}\left[1 - \sum_{g=1}^{L} \varphi_{fg}\left(1 - \overline{Q}_{fg}^{P}\right)\overline{P}_{rgf}^{\Pi B}\left(1 - \overline{\rho}_{kl}^{k}\right)\right], \qquad (2.16)$$

where Q_{jhkl}, $Q_{kg}^{P}s$—probability that the kth software module stored in the lth node will be destroyed before being referenced by the hth subscriber of the jth element, and probability of destruction of the backup of the kth software module stored in the gth node, respectively:

$$Q_{jhkl} = r_{kl} + (1 - r_{kl})g_{jhkl},$$
$$Q_{kg}^{P} = \left[r_{kg} + (1 - r_{kg})\rho_{kd}\right]Z_{k}; \qquad (2.17)$$

$\overline{Q}_{lkfr}$, $\overline{Q}_{fg}^{P}$—probability that the fth information array stored in the rth node will be destroyed before being referenced by the kth software module stored on the lth element, and probability of destruction of the backup of the fth information array stored on the gth node, respectively:

$$\overline{Q}_{1kfr} = \bar{r}_{fr} + (1 - \bar{r}_{kl})\bar{g}_{1kfr},$$
$$\overline{Q}_{fg}^{P} = \left[\bar{r}_{fg} + (1 - \bar{r}_{fg})\overline{\rho}_{fg}\right]Z_{f}; \qquad (2.18)$$

$P_{ijk}^{\Pi B}$—probability of successful transfer of a copy of the kth software module from the jth node to the ith node:

$$P_{ijk}^{n6} = P_{ijk}^{36} \left[P_{jik}^{om6} \right]^{C_k} ;$$

$\overline{P}_{rgf}^{\Pi B}$—probability of successful transfer of a copy of the fth information array from the gth node to the rth node:

$$\overline{P}_{rgf}^{\,n6} = \overline{P}_{rgf}^{\,36} \left[\overline{P}_{grf}^{\,om6} \right]^{\bar{C}_r} ;$$

$C_k(\overline{C}_f)$—number of messages containing the kth software module (the fth data array);

$P_{ijk}^{3B} (\overline{P}_{rgf}^{3B})$—probability of communicating the request for the restoration of the kth software module (fth information array) from the ith (gth) node to the jth (rth) node;

$P_{ijk}^{OTB} (\overline{P}_{grf}^{OTB})$—probability of communicating the message containing the part of the kth software module (fth information array) from the ith (rth) node to the ith (gth) node;

t_{kj}^{B}—average recovery time of the kth software module in the jth node:

$$t_{kj}^{B} = \sum_{g=1}^{L} \psi_{kg} \left[T_{jgr}^{3B} + T_{gjk}^{CB} C_k + \sum_{n=0}^{z_k-1} \left[r_{kg} + \left(1 - r_{kg}\right)\rho_{kg} \right]^n (n+1)\tau_{kg}^{B} \right];$$

$\bar{t}_{fr}^{B}$—average recovery time of the fth information array in the rth node:

$$\bar{t}_{fr}^{B} = \sum_{g=1}^{L} \varphi_{rg} \left[T_{rgf}^{36} + T_{grf}^{c6} C_k + \sum_{n=0}^{\bar{z}_f-1} \left[\overline{r}_{fg} + \left(1-\overline{r}_{fg} \right) \rho_{fg} \right]^n (n+1)\bar{\tau}_{kg}^{6} \right];$$

$T_{jgk}^{3B} (\overline{T}_{rgf}^{3B})$—average time to communicate the request for the restoration of the kth software module (fth information array) from the ith (rth) node to the gth node;

$T_{gjk}^{cB} (\overline{T}_{grf}^{cB})$—average time to communicate the message containing the part of the kth software module (fth information array) from the gth node to the rth node;

$\tau_{kg}^{B} (\overline{\tau}_{fg}^{B})$—time to produce the copy of the kth software module (fth information array) in node g;

$r_{kl} (\overline{r}_{fg}^{B})$—probability of failure of the kth software module (fth information array) when stored on the lth (gth) node until its use;

$g_{jhkl} (\overline{g}_{lkfr})$—probability of destruction of the kth software module (fth information array) distributed to the lth (r-th) node, when referenced by the hth subscriber (kth software module) of the jth node (located on the jth node);

$\rho_{kg} (\overline{\rho}_{fg})$—probability of destruction of the copy (pre-history) of the kth software module (fth information array) in the gth network node during the recovery process;

$\rho_{jl}^k (\overline{\rho}_{fg}^k)$—probability of failure of the restored copy of the jth software module (fth information array) when restored in node $l(g)$;

F_{ij}—number of information messages transmitted during transmission from the ith network node to the jth node;

$T_{ijhk} (\overline{T}_{lrkf})$—average time to transfer the message from the ith (lth) node to the jth (rth) node in the solution (reference) to the kth problem (reference to the fth information array) by the hth subscriber (reference from the kth software module) of the jth element (located in the lth node):

$$T_{jihk} = T_{jihk}^3 + T_{ijhk}^c$$
$$\overline{T}_{lrkf} = \overline{T}_{lrkf}^3 + \overline{T}_{rlkf}^c;$$

$T_{jihk}^3 (\overline{T}_{lrkf}^3)$, $T_{ijhk}^C (\overline{T}_{rlkf}^C)$—average time to communicate a solution request (reference) and a message containing solution (reference) results of the kth problem (reference to the fth information array) by the hth subscriber (reference from the kth software module) of the jth element (located in the lth node) in the ith network node (located in the rth node), respectively [17];

q_{jhkf}—number of references from the kth software module to the fth information array when solved by the hth subscriber of the jth element;

$t_{jhkl}^{\text{peш}}$—time to solve the kth software module to the lth element by the hth subscriber of the jth node in the presence of all the input data;

$\tau_{jhk} = 1$, if the hth subscriber of the jth element has the rights to solve the kth problem, $\tau_{jhk} = 0$, otherwise;

λ_{jhk}—intensity of the solution to the kth problem by the hth subscriber of the jth element;

$l_{jhk}^3 (\overline{l}_{kgf}^3)$—length of the request for the solution to (reference to) the kth problem (fth information array) by the hth subscriber (kth software module) of the jth element (with the gth reference to it);

$l_{jhk}^C (\overline{l}_{kgf}^C)$—length of the message received as a result of the solution of (reference to) the kth software module (fth information array) by the hth subscriber (kth software module) of the jth element (with the gth reference to it);

$l_k^B (\overline{l}_f^B)$—length of the request to restore the kth software module (fth information array);

u_k—volume of the kth software module;

δ_f—volume of the fth information array.

The following criteria can be used in the formulation of network information CR optimization problems in the network: maximum probability of solving all problems, minimum time for solving all problems, and minimum amount of information circulating in the network.

As a result of solving each problem, it is necessary to determine a subset of network nodes, while the placement of a program module (information array) and their backup in each of them ensures the extreme value of the optimization criterion used. In addition, when solving optimization problems of a computing resource by the criteria of maximum probability of solving all problems and the minimum time to solve them, it is necessary to determine the backup amount.

The problems of optimizing the computing resource of information by each of the listed criteria are given as follows.

1.   Determine values x_{kj}, y_{fj}, ψ_{kj}, φ_{fj}, z_k, $\bar{z}_f$ ($k = 1, 2, \ldots, K; j = 1, 2, \ldots, L; f = 1, 2, \ldots, M$) such that

$$P = \max \prod_{j=1}^{L} \prod_{h=1}^{m_j} \prod_{k=1}^{k} P_{jhk}(x, y, \psi, \varphi, z, \bar{z}) \qquad (2.19)$$

under the following constraints:

(a)   on the time to solve the kth problem by the hth subscriber of the jth element

$$T_{jhk}^{peu} \leq T_{jhk}^{\partial on} , j = 1,2,\ldots,L; h = 1,2,\ldots,m_j; k = 1,2,\ldots,M; \qquad (2.20)$$

(b)   on the amount of information circulating in the network when the hth subscriber of the jth node solves the kth problem

$$\Lambda_{jhk} \leq \Lambda_{jhk}^{\text{доп}}, j = 1,2,\ldots,L; h = 1,2,\ldots,m_j; k = 1,2,\ldots,M; \qquad (2.21)$$

(c)   on the volume of the external storage device of the jth network element

$$\sum_{k=1}^{K} (x_{kj} + y_{kj}z_k)u_k + \sum_{f=1}^{M} (y_{fj} + \varphi_{fj}\bar{z}_f)\delta_f \leq V_j, j = 1, 2, \ldots, L; \qquad (2.22)$$

(d)   on the values of variables

$$\sum_{j=1}^{L} x_{kj} = 1, k = 1, 2, \ldots, K; \sum_{j=1}^{L} y_{gj} = 1, g = 1, 2, \ldots, M, \qquad (2.23)$$

$$x_{kj} = \{0, 1\}; y_{gj} = \{0, 1\}; k = 1, 2, \ldots, K; g = 1, 2, \ldots, M; j = 1, 2, \ldots, L, \qquad (2.24)$$

$$\sum_{j=1}^{L} \psi_{kj} = 1, \ k = 1, 2, \ldots, K; \ \sum_{j=1}^{L} \varphi_{gj} = 1, g = 1, 2, \ldots, M, \quad (2.25)$$

$$\psi_{kj} = \{0, 1\}; \ \varphi_{gj} = \{0, 1\}; \ k = 1, 2, \ldots, K; g = 1, 2, \ldots, M; j = 1, 2, \ldots, L, \quad (2.26)$$

$$x_{kj} + \psi_{kj} < 2; y_{fj} + \varphi_{fj} < 2, k = 1, 2, \ldots, K; \ f = 1, 2, \ldots, M; j = 1, 2, \ldots, L, \quad (2.27)$$

$$z_k, z_f = (0, 1, 2, 3, \ldots) \ (k = 1, 2, \ldots, K; \ g, f = 1, 2, \ldots, M), \quad (2.28)$$

where V_j—volume of the external storage device of the jth network element,
T_{jhk}^{don}—maximum time for the hth subscriber of the jth computer to solve the kth problem, and
Λ_{jhk}^{don}—maximum allowable amount of information circulating in the network when the hth subscriber of the jth computer solves the kth problem.

2. Determine values

$$x_{kj}, y_{fj}, \psi_{kj}, \varphi_{fj}, z_k, \bar{z}_f (k = 1, 2, \ldots, K; j = 1, 2, \ldots, L; f = 1, 2, \ldots, M),$$

such that

$$T = min \sum_{j=1}^{L} \sum_{h=1}^{m_j} \sum_{k=1}^{K} T_{jhk}^{peu} \left(x, y, \varphi, \psi, z, \bar{z} \right) \quad (2.29)$$

under constraints (2.21–2.28) and the probability of solution to the kth problem by the hth subscriber of the jth element

$$P_{jhk} \geq P_{jhk}^{\text{доп}}, j = 1,2,\ldots,L, h = 1,2,\ldots,m_j, k = 1,2,\ldots,M, \quad (2.30)$$

where—minimum allowable value of the probability of solving all problems in the network.

3. Determine values

$$x_{kj}, y_{fj}, \psi_{kj}, \varphi_{fj}, z_k = 1, \bar{z}_f = 1 \ (k = 1, 2, \ldots, K; j = 1, 2, \ldots, L; f = 1, 2, \ldots, M),$$

such that

$$\Lambda = min \sum_{j=1}^{L} \sum_{h=1}^{m_j} \sum_{k=1}^{K} \Lambda_{jhk}(x, y, \varphi, \psi) \quad (2.31)$$

under constraints (2.20), (2.22)–(2.28), (2.30).

In order to ensure more reliable operation of some network nodes, as well as the possibility of autonomous operation of individual computational tools, forced distribution of software modules and information arrays over network nodes can be provided [18].

To account for this feature, the following notations are introduced:

$B = \|b_{kj}\|$—matrix of the preferred distribution of the information array over the network nodes, $k = 1, 2, \ldots, K, j = 1, 2, \ldots, L$;

$\overline{B} = \|\overline{b}_{fj}\|$—matrix of the preferred distribution of the information array over the network nodes, $k = 1, 2, \ldots, K, f = 1, 2, \ldots, M$, where

$$b_{kj} = \begin{cases} 1, \text{ if } k \text{ program} \ldots \text{ module is intended to be placed on } j \text{ element,} \\ 0 - \text{otherwise;} \end{cases}$$

$$\overline{b}_{fj} = \begin{cases} 1, \text{ if f information array is intended to be placed on } j \text{ element,} \\ 0 - \text{otherwise.} \end{cases}$$

Then, when the subscribers of the jth node solve the kth problem (reference from the kth software module to the fth information array), if $b_{kj} = 1$ $(\overline{b}_{fj} = 1)$, message transmission time $T_{jihk}(\overline{T}_{lrkf})$ is zero, probability of information transfer $P^{\Pi}_{jihk}(\overline{P}^{\Pi}_{lrkf})$—unity, and expression (2.22) is converted to the form

$$\sum_{k=1}^{K} (\zeta_{kj} + \psi_{kj}z_k)u_k + \sum_{f=1}^{M} (\xi_{fj} + \varphi_{fj}z_f)\delta_f \leq V_j, j = 1, 2, \ldots, L,$$

where

$$\zeta_{kj} = x_{kj} \vee b_{kj};$$

$$\xi_{fj} = y_{fj} \vee b_{fj}.$$

In addition, the following constraints will be added to the specified statements of problems to optimize the computing resource in the network, which are imposed on the number of software modules and information arrays distributed to the jth network element.

$$\sum_{k=1}^{K} x_{kj} \geq \sum_{k=1}^{K} b_{kj},$$

$$\sum_{m=1}^{M} y_{mj} \geq \sum_{m=1}^{M} \overline{b}_{mj}, j = 1, 2, \ldots, L.$$

Due to the complexity of dependencies (2.12)–(2.14), a large number of variables and constraints, solution to problems (2.19)–(2.28), (2.29), (2.21)–(2.28), (2.30) and (2.31), (2.20), (2.22)–(2.28), and (2.30) by traditional methods is difficult.

To obtain an approximate solution to the problem, the algorithm below can be used.

To simplify the proposed algorithm, the set of Boolean variables

$$\left\{ x_{kj}, y_{fj}, \psi_{kj}, \varphi_{fj} \right\}$$

and integer variables

$$\left\{ z_k, \bar{z}_f \right\} (j = 1, 2, \ldots, L; k = 1, 2, \ldots, K; j = 1, 2, \ldots, M,)$$

are replaced by a set of integer variables $\{\gamma_i\}$ ($i = 1, 2, \ldots, 3(K+M)$) taking values 1, 2, …, $\gamma_i^{\max}$, while for $i = 1, 2, \ldots, K$, γ_i correspond to x_{kj}; for $i = K + 1, K + 2, \ldots, K + M$—$y_{fj}$; for $i = K + M + 1, K + M + 2, \ldots, 2K + M$—$\psi_{kj}$; for $i = 2K + M + 1, 2K + M + 2, \ldots, 2(K + M)$—$\varphi_{fj}$; for $i = 2(K + M) + 1, 2(K + M) + 2, \ldots, 3K + 2M$—$z_k$. At the same time, when $i = 1, 2, \ldots, 2(K + M)$, $\gamma_i^{\max} = L$, and with $i = 2(K + M) + 1, 2(K + M) + 2, \ldots, 3(K + M)$ values, $\gamma_i^{\max}$ represent the maximum allowable volume of recovery backup determined by the characteristics of the system.

In this case, expressions (2.12)–(2.18), (2.22), as well as dependencies to define t_{kj}^B and $\bar{t}_{fr}^B$, are transformed into the following form:

$$P_{ihk} = \tau_{jhk} P_{jhk\gamma_k}^p p_{j\gamma_k hk}^n \prod_{f=1}^{M} (P_{\gamma_{kf}\gamma_{K+f}}^0 \overline{P}_{\gamma_k\gamma_{K+f}kf}^{\Pi})^{g_{jhkf}};$$

$$T_{jhk}^{\text{рНвщ}} = \frac{1}{\tau_{jhk}} = (T_{j\gamma_k hk} + t_{jhk\gamma_k}^{\text{реш}} + Q_{jhk\gamma_k} t_{k\gamma_k}^B) +$$

$$+ \sum_{f=1}^{M} q_{jhkf} (\overline{T}_{\gamma_k\gamma_{K+f}kf} + \overline{Q}_{\gamma_k kf\gamma_{K+f}} \bar{t}_{f\gamma_{K+f}}^B),$$

$$L_{jhk} = 1_{jhk} \left[\left\{ \left[F_{j\gamma_k} 1_{jhk}^3 + F_{\gamma_{kj}} 1_{jhk}^c + Q_{jhk\gamma_k} (F_{i\gamma_{K+M+k}} 1_k^B + F_{li} u_k) \right] + \right.$$

$$+ \sum_{f=1}^{M} \sum_{g=1}^{q_{jhkf}} \left[F_{\gamma_{K+f}\gamma_k} \bar{1}_{kgf}^3 + F_{\gamma_k\gamma_{K+f}} \bar{1}_{kgf}^c + \overline{Q}_{\gamma_{kkf}\gamma_{K+f}} (F_{\gamma_{K+f}\gamma_{2K+M+f}} \bar{1}_f^B + \right.$$

$$+ F_{\gamma_{2K+M+f}\gamma_{K+F}} \delta_f) \left] \right\} \right];$$

$$P_{jhk\gamma_k}^P = 1 - Q_{jhk\gamma_k} \left[1 - (1 - Q_{k\gamma_{K+M+k}}^P) P_{v_k\gamma_{K+M+k}k}^{\Pi B} (1 - \rho_{k\gamma_k}^k) \right];$$

$$P_{\gamma_{kkf}\gamma_{K+f}}^0 = 1 - \overline{Q}_{\gamma_{kkf}\gamma_{K+f}} \left[1 - (1 - \overline{Q}_{f\gamma_{2K+M+f}}^P) \overline{P}_{\gamma_{K+f}\gamma_{2K+M+f}f}^{\Pi B} (1 - \overline{\rho}_{f\gamma_{2K+M+f}}^k) \right];$$

$$Q_{k\gamma_{K+M+k}}^P = \left[r_{k\gamma_{K+M+k}} + (1 - r_{k\gamma_{K+M+k}})\rho_{k\gamma_{K+M+k}} \right]^{\gamma_{2(K+M)+k}};$$

$$\overline{Q}^{\,P}_{f\,\gamma_{2K+M+f}} = \left[\bar{r}_{f\,\gamma_{2K+M+f}} + (1 - \bar{r}_{f\,\gamma_{2K+M+f}})\overline{P}_{f\,\gamma_{2K+M+f}}\right]^{\gamma_{3K+2M+f}};$$

$$\sum_{k=1}^{K}\left(signf(\gamma_k, j) + signf(\gamma_{K+M+k}, j)\gamma_{2(K+M)+k}\right)u_k +$$

$$+ \sum_{f=1}^{M}(signf(\gamma_{K+f}, j) + signf(\gamma_{2K+M+f}, j)\gamma_{3K+2M+f})\delta_f V_j,$$

$$t^B_{kj} = T^{36}_{j\gamma_{K+M+k}} + T^{c6}_{\gamma_{K+M+k}{}^{jk}}C_k + \sum_{n=0}^{\gamma_{2(K+M)+k}-1}\left[r_{k\gamma_{K+M+k}} + (1 - r_{k\gamma_{K+M+k}})\rho_{k\gamma_{K+M+k}}\right]^n(n+1)\tau^6_{k\gamma_{K+M+k}};$$

$$\bar{t}^B_{f\gamma_{K+f}} = \overline{T}^{3B}_{\gamma_{K+f}\gamma_{2K+M+f}} + \overline{T}^{CB}_{\gamma_{2K+M+f}\gamma_{K+f}\,f}\overline{C}_f$$

$$+ \sum_{n=0}^{\gamma_{3K+2M+f}-1}\left[\bar{r}_{f\gamma_{2K+M+f}} + (1 - \bar{r}_{f\gamma_{2K+M+f}})\overline{P}_{f\,\gamma_{2K+M+f}}\right]^n(n+1)\overline{\tau}^B_{f\,\gamma_{2K+M+f}};$$

where

$$signf(\gamma_i, j) = \begin{cases} 1, \text{if}\gamma_1 = j. \\ 0 - \text{otherwise} \end{cases} \quad l = 1, 2, \dots, 3\,(K + M).$$

In this case, an approximate algorithm for solving the problem of optimizing computational flow intensity by the criterion of the maximum probability of solving all problems can be formulated as follows.

Step 1.
Take P^0, $P^1 = 0$, γ_i^0, $\gamma_i^1 = 0$ ($\gamma_i^0 \in \Gamma^0$, $\gamma_i^1 \in \Gamma^1$), $i = 1, 2, \dots, 3(K + M)$.
Step 2. Take $i = 1$, $\bar{u} = 0$.
Step 3. At each ith step of the algorithm, perform the following actions:
Step 3.1. Determine values P, T, Λ at $\gamma_i = 1, 2, \dots, \gamma_i^{\max}$, excluding $\gamma_i = \bar{u}$.
Step 3.2. Take $\gamma_i^1 = \gamma_i$, at which parameter $P^1 = P$ takes the maximum value while satisfying all constraints. If there are no such values, that is, at any value $\gamma_i\,\Gamma^1$ does not fall within the range of acceptable values, go to step 3.3, otherwise, to step 4.
Step 3.3. Take $i = i - 1$. If $i > 0$, put $\bar{u} = \gamma_i$ and go to step 3, otherwise go to step 7.
Step 4. Take $i = i + 1$.
Step 5. If $i \leq 3(K + M)$, go to step 3.
Step 6. If $P^1 > P^0$, take $P^0 = P^1$, $\Gamma^0 = \Gamma^1$ and go to step 2, otherwise go to step 7.
Step 7. Finish the calculations. Γ^0—solution to the problem.

An approximate solution to this problem according to the criteria of the minimum time for solving all problems in the network and the minimum amount of information circulating in this case can be found by a similar scheme.

However, for systems operating in real time (in the case of distribution (redistribution) of the software module, information array, and their computing resource at the stage of operation and functioning), solving this problem within a reasonable time even by an approximate method does not seem to be possible due to the large number of variables and constraints [19].

In this regard, it is necessary to propose a decomposition-based approach to reducing the dimensionality of the general problems of optimizing the computational resource of information in the CA FNS and CS.

Thus, a general mathematical model to optimize the computational resource of information in modern CA FNS and CS has been developed. The problem of optimizing the computational resource of network information is proposed to be solved by the criterion of the maximum probability of solving all problems, the minimum time to solve all problems, and the minimum amount of information circulating in the network. As a result of solving the problem, a subset of network nodes is determined, while the placement of a program module (information array) and their backup in each of them ensures the extreme value of the optimization criterion used. In addition, when solving optimization problems of a computing resource by the criteria of maximum probability of solving all problems and the minimum time to solve them, it is necessary to determine the backup amount.

2.4.2 Consideration of the Functioning of Flight and Computer Systems in the Context of Functional Network Degradation

One of the requirements for modern CA FNS and CS is the need for their operational stability under the conditions of functional network degradation, which is typical of systems operating under extreme conditions. In this case, in the event of a significant deterioration in the aggregate characteristics of computing facilities (for example, the total capacity of information storage devices), it is necessary to ensure priority distribution (redistribution) of software modules and information arrays and their backup having the greatest importance for the system in terms of performance of its problems, among network nodes.

The following cases are possible:

(1) decrease in computing power of FNS and CS;
(2) decrease in the total storage capacity;
(3) reduction of the total number and bandwidth of data transmission channels.

To take into account this possibility, it is advisable to introduce the criticality coefficients for software modules and information arrays—β_k, $\overline{\beta_f}$ $(k = 1, 2, \ldots, K, f =$

$1, 2, \ldots, F$), respectively, such that

$$\prod_{k=1}^{K} \beta_k = 1, \qquad \prod_{f=1}^{F} \overline{\beta_f} = 1.$$

In this case, the normalizing dependence will have the following form for the criticality coefficient of the kth software module and the criticality coefficient of the fth information array, respectively,

$$\beta_k = \frac{\beta_k'}{\sqrt[K]{\prod_{k=1}^{K} \beta_k'}}; \quad \overline{\beta_f} = \frac{\overline{\beta_f'}}{\sqrt[F]{\prod_{f=1}^{F} \overline{\beta_f'}}}.$$

In this case, dependence (2.12) will take the following form:

$$P = \prod_{k=1}^{K} \beta_k \prod_{j=1}^{L} \prod_{h=1}^{m_j} \tau_{jhk} \sum_{l=1}^{L} P_{jhkl}^{p} P_{jlhk}^{\Pi} x_{kl} \prod_{f=1}^{M} \overline{\beta_f} \sum_{r=1}^{L} y_{fr} \left(P_{lkfr}^{o} P_{lrkf}^{\Pi} \right)^{q_{jhkf}}.$$

The use of this approach will make it possible, in the first place, to distribute (redistribute) software modules and information arrays that are of primary importance to the CA FNS and CS functioning processes [20].

2.4.3 Decomposition of the General Problem of Optimization of Recovery Information Backup in Modern Flight-Navigation and Computer Systems

To reduce the dimension and computational complexity of the general problem of optimizing the computational resource of information in modern and future FNS and CS, it is proposed to decompose it into three interrelated subproblems: distribution of software modules and information arrays over network nodes, distribution of recovery backup of software modules and information arrays over network nodes, and determination of the backup amount for each software module (information array).

The proposed decomposition of the general problem of optimizing the computational resource of information in the CA FNS and CS simplifies the optimization calculations, but requires the development of mathematical statements of the problems formed in the decomposition process and rational algorithms for solving them. Below are the results of studies conducted in this field.

2.4.3.1 Mathematical Model for Distributing Software Modules and Information Arrays Over the Nodes of the Network of Flight-Navigation and Computer Systems According to the Criterion of Minimum Transmitted Information

Assuming that the probability of destruction of software modules and information arrays during operation and storage is equal to zero, we will obtain expressions (2.12)–(2.14) for P_{jhk}, Λ_{jhk}, and T_{jhk}^{pem} as

$$P_{jhk} = \tau_{jhk} \sum_{l=1}^{L} x_{kl} P_{jlhk}^{\Pi} \prod_{f=1}^{M} \sum_{r=1}^{L} y_{fr} (\bar{P}_{lrkf}^{\Pi})^{q_{jhkf}},$$

$$T_{jhk}^{pem} = \sum_{l=1}^{L} x_{kl} (T_{jlhk} + t_{jhkl}^{pem}) + \sum_{f=1}^{M} q_{jhkf} \sum_{r=1}^{L} \bar{T}_{lrkf} y_{fr},$$

$$\Lambda_{jhk} = \lambda_{jhk} \left[\sum_{i=1}^{L} \left\{ (F_{ji} 1_{jhk}^{3} + F_{ij} 1_{jhk}^{0}) x_{ki} + \right. \right.$$

$$\left. \left. + \sum_{f=1}^{M} \sum_{r=1}^{L} \sum_{g=1}^{q_{jhkf}} y_{fr} (F_{ri} \bar{1}_{kgf}^{3} + F_{ir} \bar{1}_{kgf}^{0}) \right\} \right]$$

Based on the expressions obtained, the problem of distributing software modules and information arrays by the criterion of the minimum amount of information circulating in the network can be formulated as follows. Determine values x_{kj} and y_{fj} ($k = 1, 2, …, K; j = 1, 2, …, L; f = 1, 2, …, M$) such that

$$\Lambda = \min \sum_{j=1}^{L} \sum_{h=1}^{m_j} \sum_{k=1}^{K} \Lambda_{jhk}(x, y) \tag{2.32}$$

under constraints (2.9), (2.12), (2.13), (2.19) and

$$\sum_{k=1}^{K} x_{kj} u_k + \sum_{f=1}^{M} y_{fj} \delta_f \leq V_j, \ (j = 1, 2, …, L). \tag{2.33}$$

The problem refers to linear discrete programming problems with mixed constraints.

2.4.3.2 Mathematical Model for Distributing the Recovery Backup Over the Nodes of the Network of Flight-Navigation and Computer Systems According to the Criterion of the Maximum Probability of Solving All Problems

Using the results of solving the problem of the distribution of software modules and information arrays over network nodes, and assuming that the probability of failure of the recovery backup is equal to zero ($z_k, \bar{z}_f = 1, \forall k, f$), expressions (2.12), (2.17), and (2.18) can be represented as follows:

$$P_{jhk} = \sum_{l=1}^{L} P_{jhkl}^P P_{jlhk}^n x_{kl}^* \prod_{f=1}^{M} \sum_{r=1}^{L} P_{lkfr}^0 (\overline{P}_{lrkf}^n y_{fr}^*)^{q_{jhkf}};$$

$$Q_{kg}^P = r_{kg} + (1 - r_{kg})\rho_{kg};$$

$$\overline{Q}_{fg}^P = \overline{r_{fg}} + (1 - \overline{r}_{fg})\overline{\rho}_{fg},$$

where x_{kl}^* and y_{fr}^* ($k = 1, 2, ..., K; f = 1, 2, ..., M; l, g = 1, 2, ..., L$)—result of the distribution of software modules and information arrays over network nodes.

In this case, the problem of the backup distribution over the network nodes by the criterion of the maximum probability of solving all the problems can be formulated as follows.

Determine values ψ_{kj} and φ_{gj} ($k = 1, 2, ..., K; g = 1, 2, ..., M; j = 1, 2, ..., L$) such that

$$P = \max \prod_{j=1}^{L} \prod_{h=1}^{m_j} \prod_{k=1}^{K} P_{jhk}(\varphi, \psi) \tag{2.34}$$

under constraints (2.20), (2.21), (2.25), (2.26) and

$$\sum_{k=1}^{K} (x_{kj}^* + \psi_{kj})u_k + \sum_{f=1}^{M} (y_{fj}^* + \varphi_{fj})\delta_f \leq V_j, \ j = 1, 2, ..., L \tag{2.35}$$

$$x_{kj}^* + \psi_{kj} < 2; y_{fj_{fj}}^* + \varphi_{fj} < 2, k = 1, 2, ..., K; f = 1, 2, ..., M; j = 1, 2, ..., L \tag{2.36}$$

This problem is a non-linear discrete programming problem, and approximate methods can be used to solve it. To find the exact solution, it is necessary to transform target function (2.34) to a linear form.

It follows from the analysis of expressions (2.12), (2.15), (2.16), (2.23), and (2.24) that

$$\frac{P^1}{P^0} = \prod_{l=1}^{L} \prod_{k=1}^{K} \Delta_{kl} \prod_{g=1}^{M} \overline{\Delta}_{gl}, \tag{2.37}$$

where

$$\Delta_{kl} = \frac{P_{kl}^B}{P^0};$$

$$\overline{\Delta}_{kl} = \frac{\overline{P_{gl}^B}}{P^0};$$

P_{kl}^B, $(\overline{P}_{gl}^B)$—probability of solving all problems with the kth software module (kth information array) placed in the lth network node;

P^1—probability of solving all problems with the volume of the recovery backup of each software module (information array) equal to one copy (pre-history);

P^0—probability of solving all problems without a computing resource

$$(z_k, \overline{z}_f = 0, k = 1, 2, \ldots, K; \ f = 1, 2, \ldots, M).$$

Taking the logarithm of (2.37) gives

$$\ln\left(\frac{P^1}{P^0}\right) = \sum_{l=1}^{L} \sum_{k=1}^{K} \left[\ln(\Delta_{kl}) + \sum_{g=1}^{M} \ln(\overline{\Delta}_{gl}) \right].$$

In this case, taking into account the fact that the function $y = \ln(x)$ is non-decreasing, problems (2.34), (2.20), (2.21), (2.25), (2.26), (2.35), and (2.36) are transformed into the following form.

Determine values $\psi_{kj}(k = 1, 2, \ldots, K)$ and $\varphi_{gj}(g = 1, 2, \ldots, M)$ $(j = 1, 2, \ldots, L)$ such that

$$\ln\left(\frac{P(\psi, \varphi)}{P^0}\right) = \max \sum_{l=1}^{L} \left[\sum_{k=1}^{K} \psi_{kl} \ln(\Delta_{kl}) + \sum_{g=1}^{M} \varphi_{gl} \ln(\overline{\Delta}_{gl}) \right].$$

under constraints (2.20), (2.21), (2.25), (2.26), and (2.35), (2.36).

The problem refers to linear discrete programming problems and can be solved by the branch and boundary method.

2.4.3.3 Mathematical Model for Determining the Amount of Recovery Backup of Information According to the Criterion of the Maximum Probability of Solving All Problems

Using the results of solving problems (2.20), (2.23), (2.24), (2.25), (2.26), (2.27), (2.30), (2.32), (2.33), (2.35), and (2.36), expressions (2.15) and (2.16) can be represented as follows:

$$P_{jhkl}^P = 1 - Q_{jhkl}\left[1 - \sum_{g=1}^{L} \psi_{kg}^*(1 - Q_{kg}^P) P_{glk}^{n\Pi B}(1 - \rho_{kl}^k) \right];$$

$$P_{lkfr}^0 = 1 - \bar{Q}_{lkfr}\left[1 - \sum_{g=1}^{L}\varphi_{fg}^*(1 - \bar{Q}_{fg})\bar{P}_{grf}(1 - \rho_{fg}^k)^{n_\theta}\right],$$

where ψ_{kg}^* and φ_{fg}^*—results of solving the problem of the recovery backup distribution over network nodes.

Based on the expressions obtained, the problem of determining the recovery backup volume is formulated as follows.

Determine values z_k and $\bar{z}_f$ $(k = 1, 2, …, K)$ $(f = 1, 2, …, M)$ such that

$$P = \max\prod_{j=1}^{L}\prod_{h=1}^{m_j}\prod_{k=1}^{K}P_{jhk}(z, \bar{z})$$

under constraints (2.20), (2.28) and

$$\sum_{k=1}^{K}(x_{kj}^* + \psi_{kj}^*z_k)u_k + \sum_{f=1}^{M}(y_{fj}^* + \varphi_{fj}^*\bar{z}_f)\delta_f \leq V_j, \quad j = 1, 2, …, L. \quad (2.38)$$

This problem is a non-linear discrete programming problem. In order to simplify the solution, it is necessary to transform it to standard discrete programming problems.

Based on the analysis of expressions (2.12), (2.15), (2.16)-(2.18), (2.25) and using the results of solving problems of distributing software modules (information arrays) over network nodes and their recovery backup, an expression to determine the probability of solving all problems can be represented as follows:

$$P(z, \bar{z}) = \prod_{j=1}^{L}\prod_{h=1}^{m_j}\prod_{k=1}^{K}\left[P_{jhk}^{P_1}(z_k)P_{jhk}^{n_1}\prod_{f=1}^{M}(P_{kf}^{0_1}(\bar{z}_f)\overline{P}_{kf}^{n_1})^{q_{jhkf}}\right],$$

where $P_{jhk}^{P_1}(z_k)$, $P_{kf}^{0_1}(\bar{z}_f)$, $P_{jhk}^{\Pi_1}$, $\overline{P}_{kf}^{\Pi_1}$ have the same physical meaning as P_{jhkl}^P, P_{kflr}^0, P_{jlhk}^{Π}, $\overline{P}_{lrkf}^{\Pi}$, respectively, but with a known distribution of software modules and information arrays and their recovery backup over network nodes.

In this case, the problem of determining the volume of the recovery backup of the software module and information array can be divided into two subproblems: determining the volume of the recovery backup of the software module and determining the volume of the recovery backup of the information array.

The first problem is formulated as follows

Determine values z_k, $k = 1, 2, …, K$ such that

$$P(z) = \max \prod_{j=1}^{L} \prod_{h=1}^{\overset{m}{j}} \prod_{k=1}^{K} P_{jhk}^{P_1}(z_k,) P_{jhk}^{n_1} \tag{2.39}$$

under constraints (2.20), (2.21), (2.28), and (2.38).

The second problem is formulated as follows

Determine values $\bar{z}_f, f = 1, 2, ..., M$ such that

$$P(\bar{z}) = \max \prod_{j=1}^{L} \prod_{h=1}^{\overset{m}{j}} \prod_{k=1}^{K} \prod_{f=1}^{M} P_{kf}^{0_1}(\bar{z}_f) \overline{P}_{kf}^{n_1})^{q_{jhkf}} \tag{2.40}$$

under constraints (2.20), (2.21), (2.28), and (2.38).

Problems (2.20), (2.21), (2.28), (2.38), (2.39), and (2.40) are optimal redundancy problems and can be solved by solving the dynamic programming functional equations.

Thus, the problems of optimizing the distribution of software modules and information arrays, as well as their recovery backup, over network nodes, and optimizing the volume of recovery backup of software modules and information arrays formed during the decomposition of the general problem of optimizing the information computing resource in FNS and CS are reduced to the standard form of discrete programming problems, which allows using existing methods for solving them.

2.5 Selection of the Composition of Information Security Systems

The availability of information on the structure of the intensity of the computational flow of CA FNS and CS, analysis of possible threats to information, and means of neutralizing them allows modeling the attacker's actions in this system in order to select the composition of the information security complexes.

Let the FNS of known purpose and configuration have the possible attacker's goals identified and a complete list of possible threats to information and means of neutralizing them (protection tools) compiled. Security tools that neutralize a specific threat constitute a line of defense.

Let's denote the total number of information threats through M; A—set of information threat numbers; F—number of possible attacker's goals in the FNS and CS; D—set of security tool numbers that can be used in the security system; B_f—set of numbers of information threats implemented by the attacker when reaching the fth target; N_j^f—set of numbers of security tools that could potentially be used to counteract the attacker's implementation of the fth goal at the jth line of defense (for neutralizing the jth threat included in the fth goal) $(f = 1, 2, ..., F; j = 1, 2, ..., M)$.

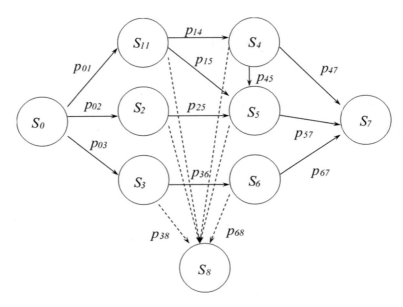

Fig. 2.2 Example of a system state graph

At the same time, $B_f \subset A$, $\bigcup\limits_{f=1}^{F} B_f = A$, $n_f = |B_f|$, and $\bigcup\limits_{f=1}^{F} \bigcup\limits_{j \in B_f}^{n_f} N_j^f \subset D$.

In this case, the process of the attacker's implementation of each of his goals can be represented as a directed graph, an example of which is shown in Fig. 2.2.

The vertices of the graph represent the state of the system corresponding to the attempt of the attacker to implement some information threat. The state of the S_0 system is initial, that is, one in which no information threat has yet been implemented. State S_j $(j \in B_f)$ corresponds to an attempt to implement the jth threat. In case of its successful implementation, the transition to the next state of the system is carried out, otherwise (with the normal response of the ISS, the security service of the system) transition to state S_{n_f+1} is carried out (in Fig. 2.2, $S_{n_f+1} \equiv S_8$).

State S_{n_f} is finite, and corresponds to the achievement of the fth attacker's goal ($f = 1, 2, ..., F$). Arcs of the graph correspond to the directions of transitions between states. Each arc is characterized by the value of the probability of transition between the corresponding states of the system. The dotted line indicates the arcs corresponding to the transition from this state to state S_{n_f+1}.

For further reasoning, it is advisable to rank the state graph. At the same time, the set of graph vertices is divided into a number of levels, which will allow the step-by-step construction of the solution to the problem of optimizing the composition of the security complex. To ensure strict sequential order of transition between the vertices of adjacent levels, it is possible to include dummy vertices into the structure of the graph. An example of state graph ranking is shown in Fig. 2.3.

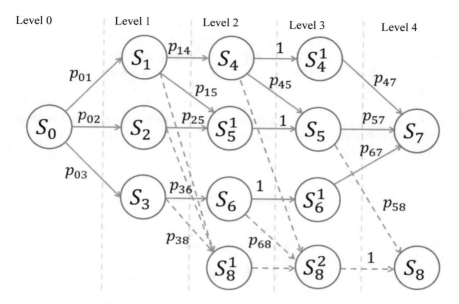

Fig. 2.3 Example of state graph ranking

Let's include dummy element 0 corresponding to system state S_0 to set A and to each of the B_f sets. In this case, the probability of finding the system in the kth state while the attacker is trying to implement the fth goal will be determined by the following expression [21]:

$$P_k^f = \sum_{l \in G_{i-1}^f} P_l^f p_{lk}^f, \quad k \in G_i^f, i = 1, 2, \ldots, I^f f = 1, 2, \ldots, F,$$

where I^f—number of levels in the ranked state graph describing the activity of the attacker when attempting to implement the fth goal; G_i^f—set of numbers of the vertices making up the ith level of the state graph describing the activities of the attacker when attempting to implement the fth goal, and

$$\bigcup_{i=0}^{I^f} G_i^f \subset B_f; \ p_{lk}^f = \rho_{lk}^f g_l^f;$$

g_j^f—probability of overcoming the jth line of defense when the attacker attempts to achieve the fth goal

$$g_j^f = \left(1 - e^{-\omega_f}\right) \prod_{m \in N_j^f} \left(1 - r_{jm}^f x_{jm}\right);$$

r_{jm}^f—probability of successful functioning of the mth security tool to counteract the attacker's activities at the jth line of defense while the attacker attempts to fulfill his fth goal ($j \in B_f; f = 1, 2, …, F; m \in N_j^f$); ω_f—mean (maximum/minimum) level of attacker's skills in the implementation of the fth goal, $\omega_f \in [0, 1]$, when the attacker attempts to implement the fth goal ($f = 1, 2, …, F$), $x_{jm} = \{0, 1\}$, $x_{jm} = 1$, if the mth tool is used at the jth line of defense, $x_{jm} = 0$ otherwise, ($j \in B_f$, $j \neq 0$, $j \neq M + 1$; $m \in N_j^f$); ρ_{lk}^f—probability of transition from the lth state of the graph to the kth one when the attacker attempts to implement the fth goal.

At the same time,

$$\sum_{k \in G_{i+1}^f} p_{lk}^f = 1, l \in G_i^f, i = 1, 2, …, I^f - 1, f = 1, 2, …, F.$$

Proceeding from the above, the values of the probabilities of the attacker's implementation of information threats (system finding in each of the states) can be calculated step-by-step in I^f steps.

At the initial moment of time (at the initial step),

$$P_o = 1; \; P_j^f = 0; \; P_{n_f+1} = 0; \; (j \in B_f; \; f = 1, 2, …, F); \; x_{jm} = 0, \forall j, m.$$

When assigning security tools that neutralize the corresponding information threat, each jth vertex of the ranked graph (with the exception of dummy vertices) will have the probability of the system's transition from the specific state to state S_{n_f+1} calculated from the following expression:

$$p_{j,n_f+1} = 1 - g_j^f.$$

It's obvious that

$$\sum_{j \in G_i^f} P_j^f = 1, (i = 0, 1, 2, …, I^f; f = 1, 2, …, F).$$

Hence, at the I^fth step, regardless of whether the information security tool is assigned to each of the lines of defense or not, we obtain the following:

$$P_{n_f} + P_{n_f+1} = 1.$$

On this assumption and given that state S_{n_f+1} along with state S_{n_f} is finite, there is no need to make intermediate calculations of the values representing the probability of the system finding in state S_{n_f+1} and the corresponding dummy vertices of the ranked graph.

The effectiveness of the navigation information security system can be determined using the following parameters:

1. Average losses of FNS and CS from the implementation of all attacker's goals:

$$C^P = \sum_{f=1}^{F} \sum_{\substack{j \in B_f, \\ j \neq 0}} P_j^f c_{jf},$$

where $c_{jf} = c_{jf}^1 + c_{jf}^2 + c_{jf}^3$; $c_{jf}^1, c_{jf}^2, c_{jf}^3$—amount of system losses from breach of information confidentiality, volume of losses from the failure to perform a number of tasks, and cost of restoring the security system if the attacker implements the jth threat while attempting to achieve the fth goal, respectively.

2. Probability of achievement of all attacker's goals:

$$P^P = \prod_{f=1}^{F} \sum_{j \in G_{tf-1}^f} P_j^f p_{jn_f}.$$

3. Probability of a successful counteraction of the security system to the attacker's actions to implement all of his goals:

$$P^3 = \prod_{f=1}^{F} \left(1 - \sum_{j \in G_{tf-1}^f} P_j^f p_{jn_f} \right).$$

4. Total cost of the security system:

$$C^3 = \sum_{f=1}^{F} \sum_{j \in B_f} \sum_{m \in N_j^f} c_{jm}^3 x_{jm},$$

where $c_{jm}^3 = s_m^0 + s_{jm}^1$—cost of using the mth tool at the jth line of defense, s_m^0—cost of the mth tool, and s_{jm}^1—cost of installation and maintenance of the mth tool at the jth line of defense.

Thus, the developed mathematical model of the attacker's actions makes it possible to evaluate the effectiveness of the system and can serve as the basis for building models for optimizing the composition of the complexes of information security tools in FNS and CS.

2.6 Mathematical Models for Optimizing the Composition of Information Security Systems

In accordance with the goals of the security system, the problem of determining the optimal composition of the complex of security tools can be solved by the criterion of the minimum probability of the achievement of all the attacker's goals, the minimum of the average losses of the system due to the attacker's implementation of all the goals, and the maximum probability of successful counteractions of the security system to the implementation of all attacker's goals. According to the first and second criteria, it is advisable to solve the problem in the case when the main goal of the security system is the maximum possible reduction in the level of implementation (predominance of software and hardware security tools). According to the third criterion, when the main problem of the security system consists in the maximum possible level of successful recognition of the fact of intrusion or actions of the attacker in the FNS with the aim of taking further measures to counter it (it is advisable to use this criterion with the predominance of physical security tools). In addition, this problem in some cases can be solved by the criterion of the minimum of the "cost–risk" integral indicator, which represents the total costs of organizing information security and the corresponding losses due to the attacker's actions. According to the above criteria, mathematical models for optimizing the composition of the complex of information security tools can be formulated as follows:

1. According to the criterion of the minimum probability of the attacker achieving all his goals, determine such values x_{jm} ($j \in B_f,\ j \neq 0;\ m \in N_j^f;\ f = 1, 2, \ldots,$ F) that

$$P^p = P^p(X) = \min_X \prod_{f=1}^{F} \sum_{j \in G_{If-1}^f} P_j^f\, p_{jn_f} \qquad (2.41)$$

under the following constraints:

$$C^3 \leq C_{\text{доп}}^3, \qquad (2.42)$$

$$\sum_{j \in G_{If-1}^f} P_j^f\, p_{jn_f} \leq P_{f\,don}^p, f = 1,2,\ldots,F, \qquad (2.43)$$

$$x_{jm} = \{0, 1\},\ (j \in B_f,\ j \neq 0;\ m \in N_j^f;\ f = 1, 2, \ldots, F), \qquad (2.44)$$

where—maximum allowable value of the cost of the security system.

2. According to the criterion of the minimum average level of system losses due to the attacker's actions, determine such values x_{jm} ($j \in B_f,\ j \neq 0;\ m \in N_j^f;\ f = 1, 2, \ldots, F$) that

$$C^p = C^p(X) = \min_X \sum_{f=1}^{F} \sum_{\substack{j \in B_f, \\ j \neq 0}} P_j^f c_{jf} \qquad (2.45)$$

under constraints (2.42)–(2.44).

3. According to the criterion of the maximum probability of successful counteraction of the security system to the attacker's actions, determine such values x_{jm} $(j \in B_f, j \neq 0,; m \in N_j^f; f = 1, 2, …, F)$ that

$$P^3 = P^3(X) = \max_X \prod_{f=1}^{F} \left(1 - \sum_{j \in G_{1f-1}^f} P_j^f p_{jn_f} \right) \qquad (2.46)$$

under constraints (2.42)–(2.44).

4. According to the criterion of the minimum of integral indicator, named "cost–risk", to determine such values x_{jm} $(j \in B_f, j \neq 0; m \in N_j^f; f = 1, 2, …, F)$ that

$$S = S(X) = \min_X \left(= \sum_{f=1}^{F} \sum_{j \in B_f} \sum_{m \in N_j^f} c_{jm}^3 x_{jm} + \sum_{f=1}^{F} \sum_{\substack{j \in B_f, \\ j \neq 0}} P_j^f c_{jf} \right) \qquad (2.47)$$

under constraints (2.43), (2.44),

where S—value of the "cost–risk" integral indicator, and $P_{f\partial on}^p$—permissible value of the probability of the attacker's implementation of the fth goal.

For the purpose of the protection system, we propose to solve the task of determining the optimal composition of the information protection system for FNS and CS with respect to the maximum probability of successful countering of the protection system to the intruder's goal accomplishment.

Tasks (2.41)–(2.44), (2.42)–(2.44), (2.45), (2.42)–(2.44), (2.46), (2.43), (2.44), and (2.47) relate to the tasks of optimization of composition of technical means complex. It is possible to use dynamic programming methods to solve them.

The computational experiment was conducted to test the performance of the developed mathematical models and algorithms. A system consisting of a server and a workstation connected by a communication channel was assumed as a system under protection.

In case the distance between the FNS and CS is insignificant, the information in the system is transmitted through a cable connection, otherwise using the radio data channel.

The analysis of the principles of the protected system functioning, the list of tasks to be solved, and the peculiarities of storage, processing, and transfer of information allowed us to identify four possible targets of the intruder, the list and brief description of which are given in Table A.1 of Appendix A. On the basis of the accumulated statistical data on the activity of intruders, analysis of possible ways of implementation of the allocated purposes by them, the list of possible threats of the information in the considered system is made, which is given in Table A.2 of Appendix A. This list is notable for its wide range and variety of threats. The greatest danger, from the point of view of the damage caused, are the threats associated with the disruption of the system's functioning (blocking of information, violation of its integrity, etc.), which requires more focus on their neutralization.

On the basis of the analysis of available statistical data, characteristics of the allocated threats of the information, the columns of conditions of the protected system are constructed at implementation by the intruder of each of the purposes presented in Figs. A.1, A.2, A.3 and A.4 of Appendix A from which it is visible that there is a considerable quantity of ways of realization by the intruder of each of the purposes that makes practically impossible the decision of a problem of definition of the structure of a complex of means of protection of the information without use of means of computing techniques [22].

The analysis of ways of implementation of the allocated threats has allowed making the list of means and the methods of protection of the information potentially suitable for inclusion in a protection system. The list of means and methods—applicants for inclusion in a protection system—is given in Table A.3 of Appendix A. The characteristics of the means of protection on the basis of which they were selected were the cost of the means and the probability of successful functioning to neutralize the relevant threat of information.

The algorithm and the program of the solution to the problem of optimization of the information protection system by the method of counter solution of functional equations of dynamic programming in Borland C++ v.5.0 language have been developed. The solution to the problem was performed for different values of limitation on the cost of protection system under OC MS Windows 98 SE control, on the aircraft with the following characteristics:

- Processor—Intel Celeron 400 MHz;
- Chip set—Intel 440 BX;
- RAM—32 MB;
- Hard disk capacity—10 GB.

The solution time was from 20 s for the lowest values to 5 min 10 s for the maximum value of the protection system cost.

The results of the solution to the problem of optimization of the information security complex composition are presented in Tables A.4, A.5 and A.6 of Appendix A and Figs. 2.4, 2.5 and 2.6.

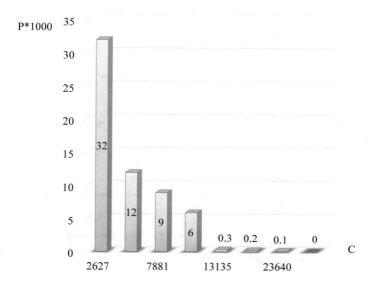

Fig. 2.4 Dependence of probability of implementation of all objectives by the intruder on the cost of the protection system

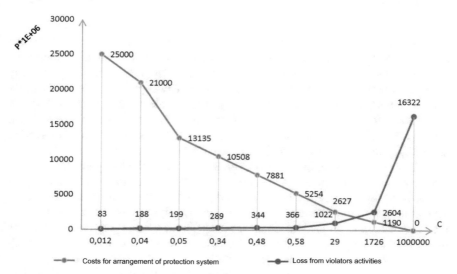

Fig. 2.5 Dependence of the expenses for the organization of the protection system and the amount of losses from the actions of the intruder on the probability of the implementation of all the goals by the intruder

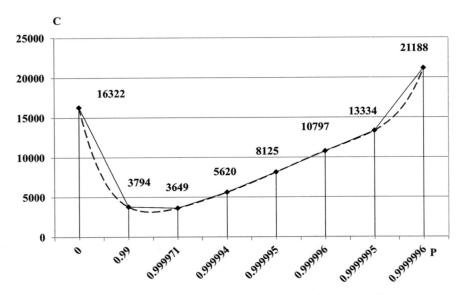

Fig. 2.6 Dependence of the total system costs on the probability of successful functioning of the protection system

Table A.4 of Appendix A and Fig. 2.4 summarize the overall results of the modeling exercise, which show that, as the amount of protection funding increases, the likelihood that an intruder will achieve all of his/her objectives is significantly reduced. Moreover, this dependence has an obviously expressed exponential character with a negative coefficient. At the same time, Table A.4 of Appendix A shows that the system is least protected against the implementation of goals Nos. 2 and 4 by the intruder, which is due to the relatively high cost of means of protection neutralizing threats 7–17 with their insufficiently high efficiency [17, 20, 21].

In Table A.5 of Appendix A and in Fig. 2.5, results of modeling are given taking into account a possibility of the use of the same information protection means (that is characteristic for means used in the computer techniques) for neutralization of several threats of the information simultaneously. Table A.6 of Appendix A shows the compositions of protection complexes for different values of imposed restrictions on the cost of the protection system. The numbers of protection means, which are included in the protection system for neutralization of other threats, but can be used for overlapping this one, are given in brackets. From the received data, it is clear that the satisfaction of the requirement for a decrease in the probability of successful realization by the intruder of all purposes leads to a sharp increase in the cost of the protection system. Thus, since some moment of loss from actions of the intruder, despite a considerable increase in the general cost of information protection means decreases insignificantly that speaks about the necessity of definition of the rational volume of expenses for the organization of protection of the information for the set initial conditions.

Figure 2.6 shows the dependence of the integral indicator of the total costs of the system of transmission and processing of information related to losses from the actions of the intruder and the costs of organizing the information protection system, the probability of successful counteraction to the system of protection of the actions of the intruder. From the given figure, it can be seen that the given dependence has a pronounced minimum where the costs of organizing the protection system and losses from the intruder's actions are equal. This suggests that from this point onwards, the cost of the protection system starts to exceed the level of losses due to the intruder's actions, and therefore the bulk of the integrated indicator value is the total cost of the protection means.

Thus, it is possible to draw a conclusion that the rational volume of expenses for the organization of the information protection system lies a little to the right of a point of a minimum of the given dependence. Thus, it can be seen from Fig. 2.6 that the increase in the volume of investments in the protection system by only three times compared to the corresponding value at this point reduces the amount of losses from the actions of the intruder by five times. At the same time, the further increase in the cost of the protection system no longer leads to a significant reduction of these losses. Hence, the corresponding option of construction of the structure of a complex of the protection means given in Table 6 (option No. 3) can be accepted as the base.

Results of experimental check of the developed mathematical models and algorithms of optimization of the structure of complexes of protection means have shown their working capacity and practical importance for working out recommendations and offers for the creation of new and improvement of existing information protection systems in systems of its transfer and processing [10.19].

2.7   Mathematical Model of Accounting for the Influence of Information Protection Means and Methods on the Functional Characteristics of the Protected Navigation System

Building an information protection system in modern FNS and CS is a complex problem that requires solving a whole range of scientific and technical problems. An information security system will only be effective if it is an integral part of the protected system. The information protection system is an auxiliary subsystem of FNS and CS and does not participate directly in the solution of tasks. Its organization requires certain resources, including material ones. The purpose of the information protection system is to ensure the reliable and stable functioning of FNS and CS in a harsh noise environment. On the one hand, each of the information protection means and methods used influences external or internal destructive factors neutralizing them to a certain extent, and, on the other hand, influences the parameters of FNS and CS functioning. Therefore, one of the requirements for information protection systems is as follows. The set of information protection means and methods used should not significantly degrade the characteristics of the protected system.

Let FNS perform W functional tasks in the course of functioning. These tasks can be divided into a number of elementary actions of information processing, the execution of which in a certain sequence leads to the solution of the corresponding task or the whole complex of tasks.

Let us denote through V—a set of numbers of elementary actions, v_i—a set of numbers of elementary actions which make up the ith task ($i = 1, 2, ..., W$),

$$v_i \subset V, \bigcup_{i=1}^{W} v_i = V.$$

The process of FNS functioning is characterized by a set of functional characteristics $H = \{h_e\}$, $e = 1, 2, ..., E$, where E is the number of functional characteristics. In order to meet the tactical and technical requirements of the system to be protected, the value of each eth characteristic must exceed (be less than) a certain permissible value:

$$h_e \begin{pmatrix} \leq \\ \geq \end{pmatrix} h_e^{\partial on}, \tag{2.48}$$

where—maximum permissible value eth functional characteristics of FNS and CS.

The formation of each eth functional characteristic contributes to the set of elementary actions, which is defined by the matrix $\overline{H}_e = \left\| \overline{h}_{ije} \right\|$, $i = 1, 2, ..., W$; $e = 1, 2, ..., E$; $j \in v_i$, where $\overline{h}_{ije}$ is the contribution of the jth elementary action when solving the ith problem to the formation of eth functional characteristic. Then

$$h_e = F_e(\overline{H}_e),$$

where $F_e()$ is a functional dependence of the value of the eth functional characteristic on the matrix of contributions of elementary actions in solving each of the problems in its formation.

The presence of an information protection system has an impact on the functional characteristics of FNS and CS.

Each of the information protection means influences the parameters of performance of elementary actions that can be set by the matrix $Z_{ije} = \left\| z_{mije} \right\|$, $i = 1, 2, ... W$; $j \in v_i$; $m \in D'$; $e = 1, 2, ..., E$, where D' is a set of numbers of means of protection included into the complex of information protection means, and z_{mije} is the size of change of the contribution of jth elementary action at the solution to ith problem on the formation of eth functional characteristic at inclusion into the information protection system of mth means. In this case

$$\overline{h}_{ije} = G_{ije}(\overline{h}_{ije}^0, Z_{ije}),$$

where $\bar{h}^0_{ije}$ is the contribution of the jth elementary action in the solution of the ith task to the formation of the eth functional characteristic without taking into account the influence of protection means, $G_{ije}()$ is the functional characteristic of the value of the contribution of the jth elementary action in the solution of the ith task to the formation of the eth functional characteristic of the value of this contribution without taking into account the influence of protection means and the Z_{ije} matrix.

In the simplest cases, $F_e()$ and $G_{ije}()$ function dependencies are additive (for example, when determining the time of solution of each task) or multiplicative (for example, when determining the probability of solution of each task). It is expedient to use dependence (2.48) in the form of additional restriction in mathematical models of optimization of the structure of complexes of information protection means.

The use of this mathematical model at the stage of design of FNS and CS will allow making reasonable decisions on the choice of the composition of the information protection system.

References

1. Akinshin RN, Esikov OV, Baryshnikov DY, Taburov DY (2004) Analysis of disturbing factors affecting data in DCN (distributed computing networks). News from the Tula State University, series "Vychislitelnaya Tekhnika. Informatsionnye tekhnologii. Sistemy Upravleniya" 1(3) (Tula)
2. Akinshin RN, Karpov IE, Samsonov AD (2013) Simulation and software package for evaluation of the flight safety system effectiveness. Nauchniy Vestnik Moscow State Tech Univ Civil Aviat 193(7):126–133
3. Akinshin RN, Andreyev AV, Rumyancev VL, Esikov OV (2016) Application of a genetic algorithm for selection of air traffic control system radio-technical facilities operating frequencies. Nauchniy Vestnik Moscow State Tech Univ Civil Aviat 19(5):126–187
4. Baburov VI, Bestugin AR, Ivancevich NV, Kirshina IA, Sauta OI, Filin AD, Shatrakov YG (2016) Comprehensive evaluation of navigation data with regard to expected uncertainty. Voprosy radioelektroniki (6):66–70
5. Bakulev PA, Sosnovskiy AA (2011) Radio navigation systems. Radiotehnika
6. Balyberdin VA (1987) Data processing systems evaluation and optimization. Radio i Svyaz
7. Balyberdin VA, Belevcev AM, Stepanov OA (2002) Optimization of information processes in automated systems with distributed data processing. Technologia
8. Zima V, Moldavyan A, Moldavyan N (2001) Global network technologies security. BHV
9. Kiselyov VD, Esikov OV, Kislitsyn AS, Theoretical basis for data processing process optimization and the structure of information protection means complexes in computing networks (edited by Sukharyov EM). JSC "Poligraphservis XXI vek"
10. Mihalevich VS, Volkovich VL (1982) Computing methods for complex systems research and engineering. Nauka
11. Pestryakov VB, Kuzenkov VD (1985) Estimation of the reliability characteristics boundary values of avionic switching system functional units according to the field performance data. Radio i Svyaz, 376 pp
12. Pestryakov VB (1990) Avionic equipment engineering. Sov. radio
13. Silin AI, Zakovryashin AI (1988) Automatic prediction of the control and monitoring equipment state. Energia
14. Trakhtengerc EA, Ivanilov EL, Yurkevich EV (2007) Modern computer technology for information analytics management. SINTEG, 320 pp

15. Falkov E, Shavrin S (2017) Cyber security of aircraft information and communication systems. Inf Analyt J "Radioelektronnye tekhnologii" (5)
16. Frolov SI, Goryachev NV, Tankov GV, Kochegarov II, Yurkov NK (2017) On some problems of reliability-oriented design of on-board avionic systems. In: Proceedings of the international symposium "Nadyozhnost i kachestvo", vol 1, pp 155–156
17. Khoroshevskiy VG (1987) Engineering analysis of computers and systems functioning. Radio i Svyaz
18. Shahtarin BI, Aslanov TG (2014) Average time to cycle slip in continuous and discrete automatic phase-lock. Vestnik of the Moscow State Technical University named after N.E. Bauman, no 1(52)
19. Yurkevich EV (2007) Introduction to the theory of information systems. LLC Izdatelskiy dom Tekhnologii, 272 pp
20. Yurkevich EV, Kryukova LN (2013) Problems of functional reliability adjustment of instrumentation and control devices in industrial processes. Izmeritelnaya Tekhnika, no 1, pp 19–23
21. Yurkov NK, Goryachev NV, Kuzina EA (2018) Physical basis of catastrophic malfunction in electrical avionic components and systems. In: Proceedings of the international symposium "Nadyozhnost i kachestvo", vol 1, pp 102–107
22. Yanbyh GF, Stolyarov BA (1987) Optimization of information computing systems. Radio i Svyaz

Chapter 3
Methods to Improve the Noise Immunity of Civil Aircraft Navigation Systems Using Satellite Radio Navigation Systems

3.1 Classification of Errors Related to Aircraft Positioning by Onboard and Ground-Based Navigation Devices

Requirements that are determined by the use of the SRNS for calculating aircraft coordinates are considered necessary to ensure the operability of aircraft navigation systems. The need to comply with the accuracy characteristics, as well as the reliability characteristics of aircraft navigation support should be attributed to these requirements [1]:

(1) availability understood as the probability of a radio-navigation system (RNS) functioning a priori or in the process of performing a certain task;
(2) integrity as a probability that a failure will be detected in the SRNS within a time interval that is less than or equal to that of the SRNS;
(3) continuity of service is defined at the most important intervals of time as the probability of operation of the SRNS.

The Required Navigation Performance Concept (RNP) formulates the requirements for aircraft navigation aids within a given airspace zone (AZ).

The accuracy of the navigation characteristics of all users within a certain amount of AZ is determined by the type of RNP. RNP are defined as follows:

(1) for the route;
(2) for several routes;
(3) district;
(4) for the AZ volume.

The following formula is used to estimate the noise error in the measurement of pseudo-range, which depends on the quality of the receiver and the constellation of the navigation spacecraft [2]:

D. A. Zatuchny et al., *Noise Resistance Enhancement in Aircraft Navigation and Connected Systems*, Springer Aerospace Technology, https://doi.org/10.1007/978-981-16-0630-4_3

$$\delta_{R_{\text{ш}}}^2 = \Delta^2 \left\{ \frac{K_1 \Delta f_{cc3}}{P/N_0} + \frac{K_2 \Delta f_{n4} \Delta f_{cc3}}{\left(P/N_0\right)^2} \right\}, \tag{3.1}$$

where $\Delta = 10^{-2} c$ is the duration of one element of the code for the equipment of SRNS consumers (CE), Δf_{cc3}, Δf_{n4}—the width of the bands, which are assumed to be equal to 3 and 100 Hz; K_1 and K_2—constant coefficients, which are equal to 0.25 and 0.5 taking into account incoherent processing adopted in the SRNS CE, P/N_0—the ratio of signal power to the spectral density of noise power.

On the basis of available statistical data, it was assumed in the calculations that $\frac{P}{N_0} = 45 \text{дб}$.

By placing the relevant data in (3.1), we obtain

$$\delta_{R_{\text{ш}}}^2 = \Delta^2 \left\{ \frac{K_1 \Delta f_{cc3}}{P/N_0} + \frac{K_2 \Delta f_{n4} \Delta f_{cc3}}{\left(P/N_0\right)^2} \right\} \approx 15{,}08 \, м^2.$$

The dependence of the noise error in pseudo-range measurement on the signal-to-noise ratio is shown on the graph in Fig. 3.1.

As it can be seen from Fig. 3.1, the dependence of the noise error in the measurement of pseudo-range on the signal-to-noise ratio has a nonlinear character and decreases sharply with increasing the signal-to-noise value.

The aircraft coordinates for the SRNS are determined on the basis of their calculation by pseudo ranges (PR) prior to the navigation spacecraft. The pseudo range

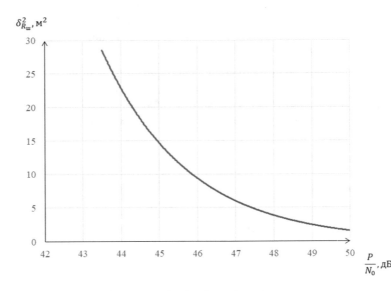

Fig. 3.1 Dependence of noise error on the signal-to-noise ratio

measured in a request-free measuring station (RFMS) is as follows [3]:$D_u(t)$

$$D_u(t) = c\tau(t) = D(t) + c\tau_\delta(t) + \delta D(t),$$

where $D(t)$ is the true distance from the RFMS to the navigation spacecraft, c is the speed of light, $\delta D(t)$ is the measurement error in the RFMS, $\tau_\delta(t)$ is the difference in the onboard time scale (OTS) compared to the single time scale (TS) of the SRNS.

Systematic errors occur when using the pseudo-dimensional method. They are caused by the difference in clock scales between the aircraft and navigation spacecraft. Thus, the determined ranges (pseudo ranges) will differ from the actual ranges by a value that is proportional to Δt, where Δt is the difference between the time scales onboard the satellite and the aircraft. This is shown in Fig. 3.2.

As it can be seen from Fig. 3.2, the difference between the true and measured range is a value proportional to Δt.

The formula used to determine pseudo range is as follows:

$$D_{iu} = \sqrt{(X - X_i)^2 + (Y - Y_i)^2 + (Z - Z_i)^2} + cT' + \delta D_i,$$

where X, Y, and Z are aircraft coordinates calculated within the geocentric coordinate system, X_i, Y_i, and Z_i are the a priori known coordinates of the ith navigation spacecraft, c is the speed of light, T' is the difference between the time scales of the navigation spacecraft and the consumer, δD_i are the errors that occur when determining the PR, $i = 1, 2, \ldots N$, N—the number of navigation spacecraft providing information for PR determination.

It should be noted that the value of T' will be equal for all D_{iu} in the assumption that all navigation spacecraft are synchronized in time with each other.

Based on the above, it becomes clear that a system of equations should be formed to determine the place and correct the time scale. Three aircraft coordinates and an error of the consumer time scale are unknown in this equation system.

From the fact that there are four unknown equations in this system, there arises the need for at least four navigation spacecraft, which will make it possible to measure four PRs. It should be noted that, as a rule, the consumer has the opportunity to use no more than 8 navigation spacecraft. Consequently, the task of determining the best constellation of four navigation spacecraft becomes relevant.

Similar nonlinear equations for pseudovelocities should be created to determine the components of aircraft speed. For pseudo-velocity calculation, we use measurements of the shifts resulting from the movement of the navigation spacecraft and the consumer based on the Doppler effect, carrier frequencies of the navigation spacecraft signals. To perform some tasks, X, Y, and Z coordinates obtained in the geocentric coordinate system are converted to the following coordinates: H-height, L-longitude, and B-latitude.

The following formulas define the relationship between these coordinates [4]:

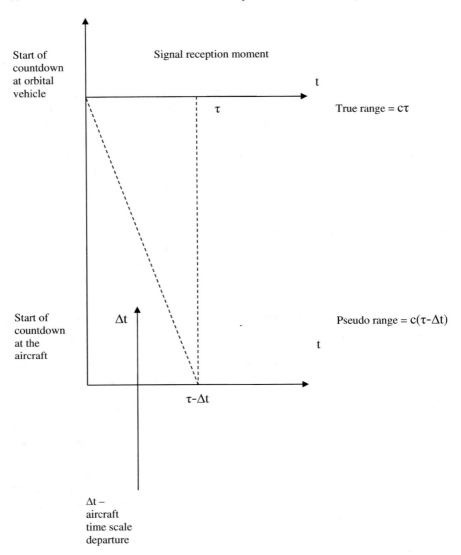

Fig. 3.2 To the notion of pseudo range

$$X = (N + H) \cos B \cos L,$$
$$Y = (N + H) \cos B \sin L,$$
$$Z = \left[(1 - e^2)N + H\right] \sin B,$$
$$N = a(1 - e^2 \sin^2 B)^{-\frac{1}{2}},$$
$$e^2 = 2\alpha - \alpha^2,$$

where α-is the compression, a is the largest half-axis of an ellipsoid.

After determining B, L, and H, the next step is the transformation of the aircraft velocity components, which becomes possible because the transition matrix is calculated from the geocentric coordinate system to a rectangular horizontal one.

The accuracy of the pseudo-range increment resulting from the measurement noise is influenced by the quality of operation of the SRNS.

The following expression allows us to determine the error of estimation of pseudo-range increment:

$$\delta_v^2 = \frac{\sqrt{2}\lambda\Delta f_{ccu}}{(2\pi)^2 (P/N_0)}. \tag{3.2}$$

It is assumed that the wavelength is $\lambda = 0{,}1\,9\,$м, and the Δf_{ccu} bandwidth for SRNS is 20 Hz.

$\dfrac{P}{N_0} = 45$дб, thus we get the following:

$$\delta_v^2 = 3{,}23\,м^2.$$

In Fig. 3.3 dependence of the error of estimation of pseudo range increment on the signal-to-noise ratio is given.

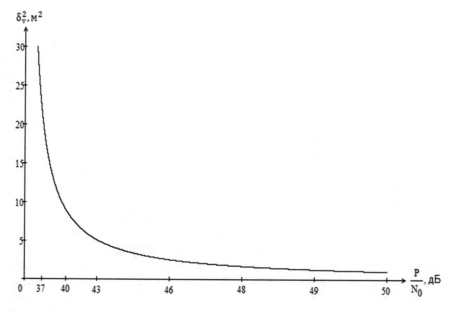

Fig. 3.3 Dependence of pseudo range increment estimation error on the signal-to-noise ratio

As it is seen from Fig. 3.3, the dependence of the error of estimation of an increment of pseudo-range from the signal-to-noise ratio has nonlinear character and sharply decreases with an increase in the signal-to-noise value. After the signal-to-noise value exceeds $50\partial\!\!6$, the noise error becomes insignificant [5].

When determining the dynamic error of measurement of coordinates, it should be divided into two types of errors. The first type includes the actual error of pseudo range measurement, as well as the error resulting from the implementation of sequential measurement of pseudo ranges up to four navigation spacecraft in time.

The following expression is used to estimate the first type of dynamic error $\delta_{R_{\text{Д}}}$ [4]:

$$\delta_{R_{\text{Д}}} = \frac{1.12\,\lambda\Delta a}{4\Delta f_{ccз}^2}, \tag{3.3}$$

where $\Delta a = 2\text{м}/c$ is the rate of change in Doppler frequency shift.

Let us insert the above values in formula (3.3). Thus we get the following:

$$\delta_{R_{\text{Д}}} = 1.06 \times 10^{-17}\,\text{м}.$$

The following expression is used to identify the second type of dynamic error δ_{R_i}:

$$\delta_{R_i} = \left(\widehat{V}_{R_i} - V_{R_i}\right)\Delta t + \left(\widehat{a}_{R_i} - a_{R_i}\right)\frac{\Delta t^2}{2}, \tag{3.4}$$

where V_{R_i}, a_{R_i} are the true values of speed and acceleration in the direction of the ith navigation spacecraft, $\widehat{V}_{R_i}$, $\widehat{a}_{R_i}$ are the velocity and acceleration estimates in the direction of the ith navigation spacecraft, Δt is the time interval from measuring the corresponding pseudo range to the moment of solving the problem of locating aircraft.

On the basis of statistical data, it is assumed that the difference between the estimation of speed and its true value is equal to $0.02\text{м}/c$, and the difference between the estimation of acceleration and its true value is equal to $0.02\,\text{м}/c^2$. Figure 3.4 shows the graph of the dependence of the dynamic error on the time interval.

As a rule, the time interval Δt sufficient to solve the navigation problem is equal to $\Delta t = 10c$. Thus, $\delta_{R_i} \approx 1.2\,\text{м}$.

In civil SRNS CE, a single frequency is usually used. Thus, compensation for ionospheric delay is based on the use of the ionospheric model. The model error is influenced by the accuracy of electron concentration accounting, which in turn is influenced by the geographical latitude, time of day and year, and the solar activity cycle phase. Corrections for ionospheric delays are determined based on geographical, daily, seasonal, and cyclic changes. This makes it possible to reduce the ionospheric error by 50–75%. The following expression gives an opportunity to estimate the residual error [3, 4]:

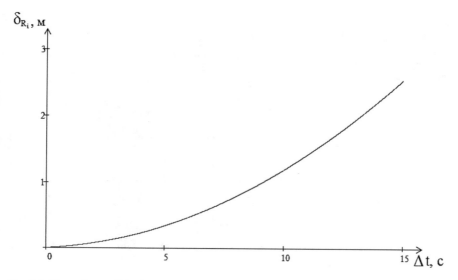

Fig. 3.4 Dependence of the dynamic error on time interval

$$\delta_{uo\text{н.}} = 2\tau_u c \cos ec\sqrt{\gamma^2 + 20^0} \approx 0.5\cos ec\sqrt{\gamma^2 + 20^0}, \tag{3.5}$$

where τ_u is the uncompensated ionospheric delay, γ is the angle of the navigation spacecraft location, and c is the speed of light. Given that the smallest angle of the navigation spacecraft location is $\gamma = 5^0$, we obtain the following: $\delta_{uo\text{н}_{min}} = 0.5\text{м}$.

The following fact should be noted. Errors resulting from the propagation specificity of the ionosphere signals are compensated for by the two-frequency equipment on the basis of the known assumption that the errors of pseudo range and the square of the carrier frequency for this case are inversely proportional.

During the propagation of radio waves in the troposphere, their velocity and trajectory are determined by the refractive index. The group delay in the troposphere has a quasidetermined character and is also compensated by using the troposphere model. The residual error $\delta_{mp.}$ is described by the expression:

$$\delta_{mp.} = \tau_m c \exp\left(-\frac{0.034}{T_0}\right)\cos ec\,\gamma \approx 0.1\exp\left(-\frac{0.034}{T_0}\right)\cos ec\,\gamma, \tag{3.6}$$

where τ_m is the uncompensated tropospheric delay (according to the accepted model), $T_0 = 300^0 K$ is the absolute temperature at the receiving point.

Given that the smallest angle of the navigation spacecraft location is $\gamma = 5^0$, we obtain the following: $\delta_{mp\cdot min} \approx 0.1\text{м}$.

Figure 3.5 shows the signal-to-noise ratio via the voltage at the transceiver's output to the angle of spacecraft position at different values of bank angles and aircraft flight

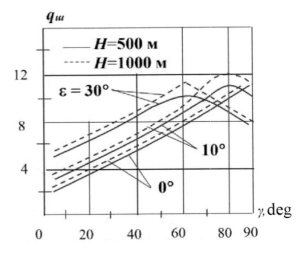

heights. It can be seen from the diagrams that at the increased flight height, the signal-to-noise ratio slightly increases, which is determined by a smaller absorption of radio waves in the atmosphere [6].

When the aircraft rolls toward the spacecraft, a typical feature is the presence of a maximum of 90°, angles $\varepsilon°$.

Figure 3.6 shows the results of the pseudo-range dependence of the RMS of pseudo range under the influence of own and external noises on the spacecraft angle.

These results are given for the aircraft flight altitude of about 1 km, when the interference caused by the influence of the underlying surface can be neglected. It

Fig. 3.6 Dependence of the root-mean-square deviation (RMS) of the pseudo-range error on the spacecraft angle under the influence of own and external noises

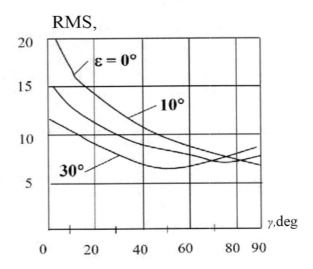

can be seen from the figure that the highest RMS value for the 5° satellite angle limitation does not exceed 20 m, and for 15°–16 m.

The calculation results of the characteristic expressed by the signal-to-noise ratio are shown in Fig. 3.7 for different location angles of the navigation aircraft (γ), aircraft flight height, and aircraft tilt angles relative to the navigation aircraft (ε).

The graphs show that the signal/interference ratio does not exceed 2 at an angle of 5° and aircraft flight at a flight altitude of up to 1 km.

The degree of impact on the value of the error in the location of aircraft noise caused by reflections from the underlying surface, the flights of aircraft at low altitudes, and the placement of the antenna on the top of the fuselage strongly depends on the angle of the location of the navigation spacecraft and the angle of the bank of the aircraft relative to the navigation spacecraft. From Fig. 3.7, it can be seen that, if the permissible angle of the aircraft roll toward the spacecraft is limited to 30°, if the angle of the spacecraft is equal to 20°, the effect of the underlying surface is manifested at altitudes less than 700 m, whereas if the angle of the spacecraft is equal to 45° at altitudes less than 100 m.

Table 3.1 shows the values of all errors.

The tropospheric correction can be calculated by means of a ratio:

$\Delta D_{Tr,i} = 8.8\mathrm{cosec}E_i$, where E_i is the angle of elevation of the ith navigation spacecraft. Application of such a correction allows you to significantly reduce the residual error.

Thus, based on the calculations presented, the following conclusion can be made: the greatest error occurs because of the noise error in the pseudo range measurement [7].

The power of the signal reflected from the underlying surface is influenced by the nature of the surface. Beckmann's model is one of the main mathematical surface models used to calculate the characteristics of the reflected signal. According to the model, the surface is rough, consisting of randomly oriented facets.

Each facet of the surface is a "shiny point" mirroring the incident wave. At that, the reflection coefficient is assumed to be equal to the Fresnel reflection coefficient. It is believed that the size of the facet is much larger than the wavelength of the

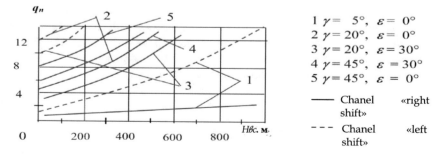

Fig. 3.7 Dependence of the signal/interference ratio on flight altitude (taking into account the influence of reflections from the Earth)

Table 3.1 Error values

Error name	Error value
Noise error in pseudo range measurement	$15.08\,\text{м}^2$
Accuracy of pseudo range increment estimation	3.23м^2
First type of dynamic error	$1.06\times10^{-17}\,\text{м}$
Second type of dynamic error	$1.2\,\text{м}$
Residual error due to ionospheric delay	0.5м
Residual error due to tropospheric delay	0.1м

incident wave ($l \gg \lambda$). The input signal will be the total of waves reflected by facets with similar orientation. The density of probabilities of distribution of the roughness heights is assumed to be normal.

Forestlands can be considered as one of the most common types of the underlying surface. Currently, there is no single mathematical model that can describe all types of forests. The following models of forests were built: thick leafless forest; rare leafless forestland (Teyk model); Klaps model describing shrubs; various models of dense deciduous forest; Teyk model for coniferous forest; snow-covered dense winter forest model [8].

With the Teyk model, the rare, coniferous forest should be regarded as a set of long and thin dielectric cylinders. These cylinders are oriented perpendicular to the plane of their location, and their distribution in this plane is random. This model belongs to the class of geometric models describing the underlying surface with roughness and is used to calculate the effective normalized areas of vegetation scattering, forest areas, etc. The following conditions are imposed in this model:

1. All cylinders are infinite in length. The scattering is usually determined by the upper part of the forest cover and does not depend on the surface underneath it.
2. The diameter of the cylinders $\left(d_u\right)$ is small compared to the wavelength. Since the main scattering occurs on tree crown branches with diameters of about 0.52 cm, this condition is fulfilled for the considered VHF range. Densities of probabilities of orientation of cylinders relative to an angle of an inclination and an angle of rotation are described by expressions:

$$W(\varphi_{ci}) = \frac{\pi}{2}\sin\varphi_{ci},$$

$$W(\theta_{ci}) = \frac{3}{2}\pi\cos^2\theta_{ci},$$

where φ_{ci} is the angle of inclination of the cylinders, and θ_{ci} is the angle of rotation of the cylinders.

The Teyk model is shown in Fig. 3.8.

The effective signal scattering area is as follows:

$$S_{эф} = \frac{1 + \cos^2 \theta_{ci}}{2} S_{эф}{}^{\Gamma} + \frac{\sin^2 \theta_{ci}}{2} S_{эф}{}^{B}, \qquad (3.7)$$

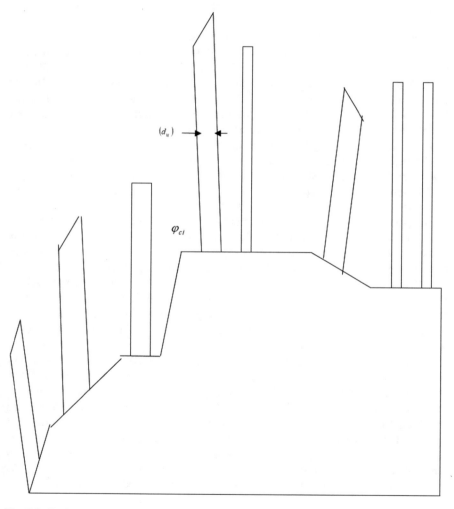

Fig. 3.8 To the explanation of Teyk model

where $S_{эф}{}^{Г}, S_{эф}{}^{B}$ is the effective scattering area for the horizontal and vertical components.

As an electromagnetic wave passes through, its amplitude fades exponentially as it penetrates inland.

Forest cover often cannot be described by one particular surface model. Large-scale model with sufficient accuracy is able to describe only the tree crowns covered with snow caps. Coniferous and leaf forests with closed tops will be complex surfaces combining large and small-scale models. The effective scattering area of a complex model can be represented by the sum of the effective scattering areas computed from private models.

Deciduous forest can be described jointly by large-scale and small-scale statistical models. At the same time, the large-scale model describes tree crowns from RMS of roughness heights of about 2 m and correlation interval of about 5 m. The small-scale model will describe the roughness caused by the foliage covering the trees. RMS of the roughness heights will be taken as 5 cm, and the interval of correlation is 8 cm. Note that the RMS of roughness heights and correlation interval in these models are mainly determined by the size of tree crowns and leaf sizes.

Radio wave scattering from coniferous forest can be described by a joint Teyk model and a small-scale model. The Teyk model describes the scattering from the trunks of coniferous forest trees and the small-scale model from branches.

The greatest level of interference is the bare terrain, and the lowest level is the bare rare forest, which can be explained by the fact that most of the electromagnetic waves penetrate the forest, where it is absorbed by multiple re-reflections Fig. 3.9 shows the change in the signal/interference ratio when flying over rugged terrain and bare, rare forest at two altitudes: $H = 1000м$ and $H = 5000м$. Line 1 shows this dependency for $H = 5000м$ height. Line 2 shows this dependency for $H = 1000м$ height. The area to the left of the shaded line shows the change in the signal-to-noise ratio on the rough terrain depending on the altitude of the aircraft flight, and the area to the right of the shaded line shows this change in a bare, rare forest. A forest with an average tree height of 3 m and a proportion of water in the branches mass of − 0.7 was considered as a bare, rare forest.

The rest of the forest cover occupies the intermediate position. Figure 3.10 shows the dependence of the signal-to-noise ratio on the aircraft flight height when flying over the coniferous forest and flying over winter forest with "caps" of snow. A coniferous forest was considered to be a coniferous forestland, with an average tree height of $\approx$ 3 m and a percentage of water content in the branches mass of $\approx$ 70%. The winter forestland was considered to be a winter forest, with an average tree height of $\approx$ 3 m. The linear dependence, which is represented by line 1, is decreasing and can be considered a characteristic at the aircraft altitude $H = 5000м$ of the signal/interference ratio. The linear dependence, which is shown in line 2, is

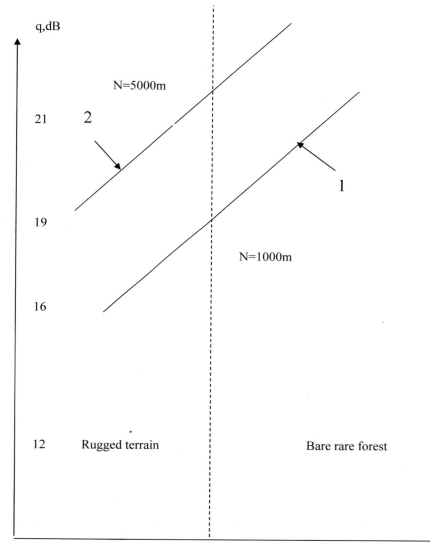

Fig. 3.9 Change in the signal/interference ratio q when flying over rugged terrain and bare rare forest

increasing and can be considered a characteristic of the signal/interference ratio at aircraft altitude of $H = 1000м$. The lines to the left of the shaded line characterize data that are relevant to coniferous forests. The lines to the right of the shaded line represent data that are relevant to the winter forest.

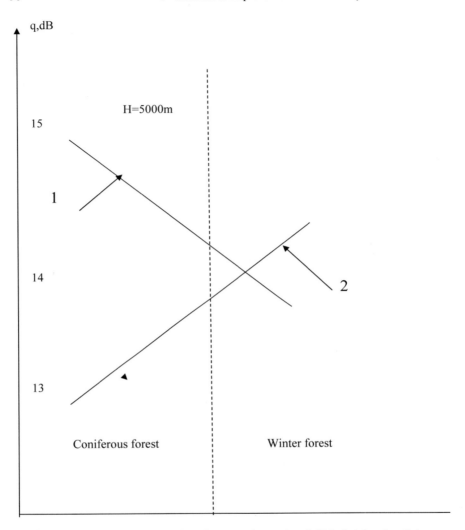

Fig. 3.10 Dependence of the signal/interference ratio on aircraft flight height when flying over coniferous forest and flying over winter forest

3.2   Working Constellation Selection Algorithm of Navigation Spacecraft in Excessive Data Load Conditions

Integrated (complex) use of SRNS GPS and GLONASS navigation signals is considered an important endeavor for the improvement of AC position location accuracy.

Table 3.2 The influence of the number of navigation spacecraft on the geometric factors

Number of navigation spacecraft	HDOP	VDOP	TDOP
8 GLONASS navigation spacecraft	1.03	1.34	0.80
10 GLONASS navigation spacecraft	0.84	1.24	0.72
All navigation spacecraft GLONASS + GPS	0.58	0.84	–

Table 3.3 Coordinates and altitude calculation accuracy using SRNS GLONASS and GPS

Mode	Coordinates; drms; m	Altitude; RMS deviation; m
All GLONASS navigation spacecraft, nom*	9.0	13.5
16 GLONASS + GPS navigation spacecraft (selective access); nom	8.5	12.6
All GLONASS + GPS navigation spacecraft (selective access); nom	8.4	12.5
All GLONASS navigation spacecraft; dif*	1.8	2.6
16 GPS navigation spacecraft + GLONASS	1.36	1.9
All GLONASS + GPS navigation spacecraft (selective access); dif	1.24	1.82

The main tasks of integration (integrated use) are to improve the accuracy and reliability comprised of meeting the conditions for integrity, availability, and continuity of service during AC position determination [9].

Accuracy characteristics received with the help of GLONASS and GPS navigation signals (GLONASS + GPS) integrated use are given in Tables 3.2 and 3.3.

Analysis of the Table shows that for the differential mode, use of all navigation spacecraft comprising SRNS GLONASS and GPS gives an improvement in altitude and coordinates calculation accuracy in 1.4 and 1.6 times compared to the use of Cases 8 and 10 GLONASS navigation spacecraft, respectively.

Analysis of the Table shows that when using GLONASS navigation spacecraft in the nominal mode in integration with GPS navigation spacecraft with selective access, the accuracy is increased only by 5–10% with regard to coordinates and 8% with regard to altitude.

When using the differential mode for both SRNS, the accuracy in determining PR was considered to be 2 m. Analysis of Table 3.3 shows that in case of using the differential mode, considering the measurement data obtained via GPS gives a 1.3…1.5 times increase in the accuracy of determining coordinates and altitudes.

It should be noted that for original GPS-users, the use of SRNS GLONASS navigation signals gives approximately a three times increase in accuracy.

Table 3.4 SRNS availability evaluation at various periods of AC flight, %

SRNS or combination of various SRNS	Aircraft routing	Aircraft local flight	Aircraft non-precision approach
GPS	98.58	96.53	67.26
GPS + GLONASS	100	99.99	98.87
GPS + GLONASS	100	100	100

Table 3.5 The longest period of inactivity, min

SRNS or combination of various SRNS	Aircraft routing	Aircraft local flight	Aircraft non-precision approach
GPS	35	70	295
GPS + GLONASS	0	15	30

Tables 3.4 and 3.5 provide availability evaluations and the longest period of inactivity when using the navigation data obtained only via SRNS GPS and SRNS GPS + GLONASS.

Analysis of Table 3.4 data shows that integrated (complex) use of GPS and GLONASS provides a considerable improvement of availability. It is the most pronounced in situations of aircraft non-precision approach (from 67 to 100%).

Analysis of the data in Table 3.5 shows that integrated (complex) use of GLONASS together with GPS during the AC track route flight leads to the absence of periods of inactivity.

To improve the AC coordinates calculation accuracy, one of the most effective ways is to select the best working constellation of the SRNS navigation spacecraft. The criterion for choosing such a working constellation shall be considered the lowest possible geometric factor in the constellation.

The accuracy characteristics of the SRNS GLONASS are influenced by the geometric location of the consumer and used navigation spacecraft, as well as allowances and errors resulting from the navigation parameters calculation process. $\eta^T = \begin{bmatrix} XYZT' \end{bmatrix}$–vector of navigation parameters.

The covariance matrix R_g of errors $\Delta\eta$ is the most common characteristic of the determination accuracy η [10]:

$$R_g = \left(H_g^T R^{*-1} H_g \right)^{-1}, \tag{3.8}$$

where T is the matrix transposition operation, -1 is the inverse matrix designation, R^* is the PR determination covariance matrix of errors, H_g is the differential derivative matrix of the pseudo-range determination function collection.

When determining the planimetric and altitude coordinates of the consumer, an expression in the form of a covariance matrix is used:

$$R = \left(H_\zeta^T R^{*-1} H_\zeta\right)^{-1}. \tag{3.9}$$

H_ζ depends on two values: H_g and a matrix demonstrating the error vector trans-position $\Delta\eta$ in the geostationary coordinate system into "the north-east-normal to the tangent plane" vector T_{gp} with consideration of the variables T' and f'.

$\zeta_1 = T_{gp}\Delta\eta$ implies that $H_\zeta = H_g T_{gp}^{-1}$.

If we consider only the accuracy characteristics of the position coordinates calculation, then (3.9) shall be written in the form:

$$R_M = \left(H_{M\zeta}^T R_{\Pi\Pi}^{*-1} H_{M\zeta}\right)^{-1}, \tag{3.10}$$

where $R_{\Pi\Pi}^*$ is the covariance matrix associated with the PR determination.

In the case when the errors have a similar physical nature and are mutually inde-pendent, and also have equal variances of the summary error $\delta_{\Pi\Pi}^2$, then $R_{\Pi\Pi}^* = \delta_{\Pi\Pi}^2 I$, where I-is a unit matrix.

$$R_M = \delta_{\Pi\Pi}^2 \left(H_{M\zeta}^T H_{M\zeta}\right)^{-1}, \tag{3.11}$$

$$\Gamma = \left(H_{M\zeta}^T H_{M\zeta}\right)^{-1}, \tag{3.12}$$

where Γ is a geometric factor.

Now let's define various geometric factors. The nature of a geometric factor is in relation to the root-mean-square errors arising when calculating the coordinates to the root-mean-square errors of determining the vector of radio-navigation parameters [11].

$GDOP = (trace\ \Gamma)^{\frac{1}{2}}$ is a common geometric factor for the navigation parameters determination accuracy changing, where $trace\ \Gamma$ is the square root of the sum of numbers located on the main diagonal line of the matrix Γ.

$PDOP = (\gamma_{11} + \gamma_{22} + \gamma_{33})^{\frac{1}{2}}$—geometric factor considered when calculating the spatial coordinates,

$HDOP = (\gamma_{11} + \gamma_{22})^{\frac{1}{2}}$—geometric factor considered when determining the coordinates in the horizontal plane,

$VDOP = \gamma_{33}^{\frac{1}{2}}$—geometric factor considered when determining the altitude,

$TDOP = \gamma_{44}^{\frac{1}{2}}$—the geometric factor considered in determining the time,

where γ_{ij} are the elements of the matrix G.

Next we consider various RMS deviations [11]:

(1)   RMS deviation when determining the planimetric coordinates:

$$\delta_B = \delta_{\Pi\Pi}\gamma_{11}^{\frac{1}{2}}, \delta_L = \delta_{\Pi\Pi}\gamma_{22}^{\frac{1}{2}}.$$

(2)   RMS deviation when determining the height:

$$\delta_H = \delta_{\Pi\!\mathcal{A}} VDOP .$$

(3)   RMS deviation when determining the time:

$$\delta_T = \delta_{\Pi\!\mathcal{A}} TDOP .$$

It should be noted that the geometrical factors are functions only for the geometrical arrangement of a navigation spacecraft and the object being determined.

$$CPO = \delta_{\Pi\!\mathcal{A}} HDOP , \quad CCO = \delta_{\Pi\!\mathcal{A}} PDOP ,$$

where *CPO* is the root-mean-square radial error, *CCO* is the root-mean-square spherical error.

Now we list the basic errors in PR determination:

(1)   errors related to obtaining ephemeris data (ED);
(2)   errors related to clock error predictions calculation;
(3)   errors related to set noise and external jamming;
(4)   errors related to multi-path propagation and peculiarities of the ionospheric and tropospheric propagation of radio waves.

$$\delta_{\Pi\!\mathcal{A}} = \sqrt{\delta_{\mathfrak{z}}^2 + \delta_c^2 + \delta_{mp}^2 + \delta_{ion}^2 + \delta_{\mathit{мн}}^2 + \delta_{\mathit{ш}}^2} , \tag{3.13}$$

where $\delta_{\mathfrak{z}}^2$—errors due to ED errors, δ_c^2——errors due to synchronization, δ_{mp}^2—errors due to knowledge of the radio waves tropospheric propagation speed, δ_{ion}^2—errors due to knowledge of the radio waves ionospheric propagation speed, $\delta_{\mathit{мн}}^2$—errors due to multi-path propagation, $\delta_{\mathit{ш}}^2$—errors due to set noise and jamming.

Totally in the worst-case scenario, by applying the modern multi-channel NCE with narrow-band single-range ($\approx$ 1600 MHz) radio-navigation signals of GLONASS system, it is possible to provide operational global navigation with the following maximum error values in three-dimensional coordinates determination: 60 m in plane and 100 m in height during the periods of maximum solar activity, 30 m in plane and 50 m in height during the periods of minimal solar activity. For comparison, in GPS system, the error with a probability of 0.95 is 100 m in plane and 156 m in height. GPS navigation accuracy is 2.5 times lower than in GLONASS [12].

In order to perform the selection procedure for the working constellation, it is necessary to look through all possible combinations of the navigation spacecrafts and calculate the geometrical factor for each combination. This calculation shall be done fairly often, because as a result of the relative motion of the AC and the NS, operational configuration of the system constantly changes resulting in the situation where the working constellation ceases to be the best choice according to the criterion of the lowest possible geometric factor with the restriction to the angle of the NS relative to the consumer.

In addition, NSs comprising the working constellation may temporarily go beyond the receiving aerial pattern, which requires a fast transition to a new working constellation without a sharp decrease in the position determination accuracy. Therefore, it is necessary to reduce the amount of calculations required to select the working constellation. It is possible to simplify this procedure by excluding from the processing the NS signals with narrow elevation angles. The limiting elevation angle is usually chosen in the range of $5–10^0$. This often leads to a decrease in the geometric factor for a specific point in time and AC position, since NS with narrow angles of elevation often comprise constellation of four with the lowest geometric factor. Therefore, it is possible to use simplified methods for selecting the optimal working constellation by searching only those combinations that include NS with the broadest elevation angle.

Positioning with the help of an excess number of NSs (for all NSs with an elevation angle bigger than the limiting angle) has several advantages. During the period of incomplete system deployment or a failure of one or several NSs, in other words, in case the geometric factor of even the best four is decently sufficient, this gain may be quite significant.

In order to evaluate the positioning error when working with an excess amount of navigation spacecrafts, it is necessary to broaden the definition of the geometric factor in the face of redundant measurements. In this regard, the concept of matrices pseudoinversion shall be used. If n spacecrafts are observed, then the cosine matrix has dimensions $n \times 4$.

Maximum likelihood evaluation of the AC state vector is

$$\overline{X} = \left(H_n^T R_n^{-1} H_n\right)^{-1} H_n^T R_n^{-1} \overline{\rho}_n,$$

where $\overline{\rho}_n$ is the measurement vector, the components of which are n of the measured pseudoranges, and R_n is the covariance matrix of this vector measurement errors.

If the observations are equal and independent, then

$$\overline{X} = \left(H_n^T H_n\right)^{-1} H_n^T \overline{\rho}_n = H^+ \overline{\rho}_n,$$

where H^+—is a matrix preudoinverse to H.

Provided that the RMS deviation of the pseudo-range measurement error equals δ_p, the covariance matrix of positioning error by the least squares technique with regard to n pseudoranges is equal to

$$R_x^{(n)} = \delta_p^2 H_n^+ \left(H_n^+\right)^T.$$

Therefore, the geometric factor is determined for positioning by an excessive number of single-step measurements n:

$$\Gamma = \left\{ T_r \left[H_n^+ \left(H_n^+ \right)^T \right] \right\}^{1/2},$$

where T_r—is the spur of a matrix designation.

To determine the coordinates of AC included into the sum of components of the vector $\overline{X}$, and to calculate the geometric positioning factor, it is necessary to calculate the pseudoinverse matrix H^+ as well. Apart from that, it is necessary to calculate expressions of the form:

$$H_n^+ R_n^+ \left(H_n^+ \right)^T.$$

It should be noted that to operate using all visible NSs, it is necessary to have modern NCE equipped, which is currently provided for a smaller part of the aircraft fleet. For that reason, the working constellation selecting procedure of four navigation spacecrafts is still relevant to this day.

Apart from that, as mentioned above, the procedure for finding the geometric factor when operating with all visible NSs is quite a labor-intensive task requiring a large amount of calculations in a small amount of time and causing difficulties even for modern computing systems.

3.3 Algorithm of Selection of the Working Constellation of Navigation Spacecraft Taking into Account the Probability of Errors in Determining Its State for Each of the Satellites

This section will provide a criterion for selecting the best working constellation of the navigation spacecraft based on the probability of errors in the determination of its state for each satellite. For this purpose, the procedure for determining these values is given below.

Two indicators are used to assess the status of navigation spacecrafts, B_i and C_i. In the process of assessing the state of the navigation spacecraft in relation to these two attributes, errors of the 1st and 2nd kind may occur:

(1) The navigation spacecraft is recognized as not conforming to the above features provided that it satisfies them. The probability of this event is indicated through ε_i for B_i and through ε_i' for C_i.

(2) The navigation spacecraft is considered to be non-complying with the above criteria provided that it does not satisfy them. The probability of this event is indicated through δ_i for B_i and through δ_i' for the C_i.

Let us suggest the following values to assess the state of each navigation spacecraft:

(a) To assess the state on the basis of B_i:
  $x_i = 0$—the satellite is accepted as functionally operative;
  $x_i = 1$—the satellite has been declared inoperable;
(b) To assess the state on the basis of C_i:
  $y_i = 0$—the satellite is accepted as functionally operative,
  $y_i = 1$—the satellite has been declared inoperable.

As mentioned above, the n-th navigation spacecraft is considered suitable for inclusion in a working constellation only if two $B_n = 0$ and $C_n = 1$ conditions are met simultaneously. Since the estimates of the state of the navigation spacecraft for each of the attributes are independent, it can be assumed that the errors that occur during the process of determining the state of the navigation spacecraft arise independently of each other.

Thus, a navigation spacecraft is suitable for inclusion in a working constellation provided that the assessment of its condition does not allow it to be considered as satisfying the selection conditions:

$$P = P(B_i = 0, C_i = 1/x_i = 1, y_i = 0)$$
$$+ P(B_i = 0, C_i = 1/x_i = 1, y_i = 1)$$
$$+ P(B_i = 0, C_i = 1/x_i = 0, y_i = 0)$$
$$= \varepsilon_i \varepsilon_i' + \varepsilon_i(1 - \varepsilon_i') + (1 - \varepsilon_i)\varepsilon_i'.$$

A navigation spacecraft is not suitable for inclusion in a working constellation provided that the assessment of its condition allows it to be considered as satisfying the selection conditions:

$$P = \delta_i \delta_i' + \delta_i(1 - \varepsilon_i') + \delta_i'(1 - \varepsilon_i).$$

Let us introduce the following symbols:

$$\varepsilon_{1,2,i} = \varepsilon_i \varepsilon_i' + \varepsilon_i(1 - \varepsilon_i') + (1 - \varepsilon_i)\varepsilon_i',$$

$$\delta_{1,2,i} = \delta_i \delta_i' + \delta_i(1 - \varepsilon_i') + \delta_i'(1 - \varepsilon_i).$$

Let us denote the probability of selecting an navigation spacecraft as p_i provided that the assessment of its condition is erroneous.

Consequently, $p_i = \varepsilon_{1,2,i} = 1 - \delta_{1,2,i}$.

Let us introduce functional dependence:
$f(x_i, y_i) = 1$, if $x_i = 0$, $y_i = 1$,
$f(x_i, y_i) = 0$ in other cases.

To indicate the status of the entire working constellation, let us introduce Φ. For the working state of the working constellation of the navigation spacecraft we will consider that $\Phi = 1$. Provided that the working constellation is in a state of inoperability, we will consider that $\Phi = 0$.

The state of the working constellation will be considered as a function of $\Phi(B,C)$, where $(B, C) = ((B_1, C_1), (B_2, C_2), \ldots, (B_n, C_n))$ is a vector of the working constellation state. The dimension of this vector n is the number of navigation spacecraft in the SRNS.

Next, the concepts of path and cross-section should be introduced [13].

The pathway is the navigation spacecraft set for which the following rule is followed: if each navigation spacecraft in the set satisfies a set of attributes $B_i = 0, C_i = 1$, then $\Phi(B, C)= 1$ regardless of the state of the other navigation spacecraft that make up the system. The set of navigation spacecraft, for which the following rule is followed, will be considered as a cross–section: if each navigation spacecraft in this set satisfies one of the feature sets: $(B_i = 1, C_i = 0)$, $(B_i = 1, C_i = 1)$, $(B_i = 0, C_i = 0)$, then $\Phi(B, C)= 0$ regardless of the state of the other navigation spacecraft that make up the system.

Next, the minimum path and minimum cross section should be defined. This is essential in order to find the minimum set of satellites that will provide or disrupt the normal operation of the entire system. It is considered that the work of all navigation spacecraft in the working constellation is independent of each other. Consequently,

$$P(B, C) = \prod_{i=1}^{n} p_i^{f(B_i,C_i)} (1 - p_i)^{1-f(B_i,C_i)},$$

where $P(B, C)$ is the probability of the event that the working constellation consists of navigation spacecraft, for which the state vector (B, C) is executed.

For the entire working constellation, let us introduce a function $R(p)$ that characterizes the probability that $\Phi(B,C) = 1$.

$R(p)$ is described by the formula:

$$R(p) = E\Phi(B, C) = \sum_{X,Y} P(B, C)\Phi(B, C) = \sum_{X,Y:\Phi(X,Y)=1} \prod_{i=1}^{n} p_i^{f(B_i,C_i)} (1 - p_i)^{1-f(B_i,C_i)}, \quad (3.14)$$

where $E\Phi(B, C)$ is the mathematical expectation of a working constellation.

Consequently, for the entire working constellation, which includes n navigation spacecraft, the error probabilities $\varepsilon_{1,2,s}$ and $\delta_{1,2,s}$ are determined by formulas:

$$\varepsilon_{1,2,s} = R(\varepsilon_{1,2,1}, \varepsilon_{1,2,2,\ldots}, \varepsilon_{1,2,n}) = R(p_1, p_2, \ldots, p_n)$$

$$= \sum_{X,Y:\Phi(X,Y)=1} \prod_{i=1}^{n} p_i^{f(x_i,y_i)} (1 - p_i)^{1-f(x_i,y_i)}$$

$$= \sum_{X,Y:\Phi(X,Y)=1} \prod_{i=1}^{n} \varepsilon_{1,2,i}^{f(x_i,y_i)} (1 - \varepsilon_{1,2,i})^{1-f(x_i,y_i)} \tag{3.15}$$

$$1 - \delta_{1,2,s} = R(1 - \delta_{1,2,1}, 1 - \delta_{1,2,2}, 1 - \delta_{1,2,n}) = R(p_1, p_2, p_n)$$

$$= \sum_{X:\Phi(X)=1} \prod_{i=1}^{n} p_i^{f(x_i,y_i)} (1 - p_i)^{1-f(x_i,y_i)}$$

$$= \sum_{X:\Phi(X)=1} \prod_{i=1}^{n} (1 - \delta_{1,2,i})^{f(x_i,y_i)} \delta_{1,2,i}^{1-f(x_i,y_i)}. \tag{3.16}$$

In the asymptotic case, the errors $\varepsilon_{1,2,i}$ and $\delta_{1,2,i}$ are recorded as follows:

$$\varepsilon_{1,2,i} = t_i \cdot \varepsilon, \tag{3.17}$$

$$\delta_{1,2,i} = t_i' \cdot \delta, \tag{3.18}$$

where t_i, t_i' are some positive constants. $\varepsilon \to 0$, $\delta \to 0$.

This clearly applies to the solution of the problem, as it is considered that each navigation spacecraft is very reliable together with the control equipment and ground control system. Thus, the probability of errors for them is unlikely.

Including (3.14)–(3.18), $\varepsilon_{1,2s}$ and $\delta_{1,2,s}$ take the following form:

$$\varepsilon_{1,2,s} = C_s^1 \varepsilon^{\alpha_s} + C_s^2 \varepsilon^{\alpha_s+1} + \cdots + C_s^{n-\alpha_s+1} \varepsilon^n; \tag{3.19}$$

$$\delta_{1,2,s} = C_s'^1 \delta^{\beta_s} + C_s'^2 \delta^{\beta_s+1} + \cdots + C_s'^{n-\beta_s+1} \delta^n. \tag{3.20}$$

Next, we should define the values α_s and β_s, as well as the constants C_s^i, $C_s'^i$. Let us introduce the following symbols:

m the weight of the minimum path from the navigation spacecraft in the working constellation, i.e., the number of navigation spacecraft in this minimum path,

m' the weight of the minimum cross section of the working constellation, i.e., the number of navigation spacecraft in that minimum cross-section,

N_{min} a lot of all the minimum paths made up of navigation spacecrafts being part of the working constellation,

N_{min}' multitude of the minimum cross-sections of a working constellation.

Then

$$\alpha_s = \min_{N_{min}} m,$$

$$\beta_s = \min_{N_{min}'} m.$$

For $t_i = t_i' = 1$ case, the first two constants in expansion ε_s and δ_s for any i-th satellite are defined as follows:

$$C_s^1 = N_{\min}\alpha_s,$$

$$C_s^2 = -(n - \alpha_s) + N(\alpha_s + 1),$$

$$C_s^{/1} = N_{\min}'\beta_s,$$

$$C_s^{/2} = -(n - \beta_s) + N(\beta_s + 1),$$

where

$N_{\min}\alpha_s$ the total number of minimum paths of the working constellation consisting of α_s navigation spacecrafts,

$N_{\min}'\beta_s$ the total number of minimum cross-sections of the working constellation consisting of β_s navigation spacecrafts.

$N(\alpha_s + 1)$ the total number of all paths of the working constellation consisting of $\alpha_s + 1$ navigation spacecrafts.

$N(\beta_s + 1)$ the total number of all cross-sections of the working constellation consisting of $\beta_s + 1$ navigation spacecrafts.

Let us introduce the designation of functionality and write down its formula taking into account (3.19), (3.20):

$$
\begin{aligned}
F_s(\varepsilon, \delta) &= \varepsilon_{1,2,s} + \delta_{1,2,s} = 1 + R(\varepsilon_{1,2,1}, \varepsilon_{1,2,2}, \ldots, \varepsilon_{1,2,n}) \\
&\quad - R(1 - \delta_{1,2,1}, 1 - \delta_{1,2,2}, \ldots, 1 - \delta_{1,2,n}) \\
&= C_s^1 \varepsilon^{\alpha_s} + \cdots + C_s^{n-\alpha_s+1} \varepsilon^n + C_s^{/1} \delta^{\beta_s} + \cdots + C_s^{/n-\beta_s+1} \delta^n.
\end{aligned} \tag{3.21}
$$

Let us formulate a criterion of a choice of an optimum working constellation: the working constellation is the best if the sum of errors of definition of a condition on each of signs will be minimum, at restriction on the geometrical factor, that is the decision of a problem consists in the performance of the following set of conditions:

$$\min F_s(\varepsilon, \delta),$$

$$\Gamma_\phi \leq \gamma,$$

where γ is a certain tolerance value for the geometric factor.

This problem is further solved for the case of heterogeneous navigation spacecraft in the meaning of reliability, which arises when combining heterogeneous systems.

Only an asymptotic case is considered when $\varepsilon, \delta \to 0$ as it is assumed that the probability of incorrect determination of the state of the navigation spacecraft is small enough.

Let us formulate and justify the following result, which is of value for the choice of the working constellation:

Let us assume that $\varepsilon, \delta \to 0$, $\varepsilon = L\delta^z$, where L is any constant, $\delta \to 0, z > 1$. In this case $\min_{s \in S} F_s(\delta) = F_{s^*}(\delta)$, where s^* is the working constellation of type $\left[\frac{n+z}{z+1}\right]$ out of n.

In this case, the given dependence between errors in the assessment of the state of each navigation spacecraft, the solution of the problem is a working constellation, which is a system in which the above conditions are satisfied by at least $\left[\frac{n+z}{z+1}\right]$ from n navigation spacecraft, where $\left[\frac{n+z}{z+1}\right]$ is a whole part of the number, z is an indicator of the degree of dependence between errors in the assessment of the state of the navigation spacecraft [14].

Taking into account (3.21), $F_s(\delta)$ functionality characterizing the suitability of an arbitrary working constellation can be rewritten as follows:

$$F_s(\delta) = C_s^1 \delta^{\alpha_s z} + \vartheta(\delta^{\alpha_s z}), \alpha_s z < \beta_s,$$

$$F_s(\delta) = C_s^{/1} \delta^{\beta_s} + \vartheta(\delta^{\beta_s}), \alpha_s z > \beta_s,$$

$$F_s(\delta) = (C_s^1 + C_s^{/1}) \delta^{\alpha_s z} + \vartheta(\delta^{\alpha_s z}), \alpha_s z = \beta_s.$$

Herewith we obtain the ratio for the best working constellation:

$$F_{s^*}(\delta) = C_{s^*}^1 \delta^{z\left[\frac{n+z}{z+1}\right]} + \vartheta\left(\delta^{z\left[\frac{n+z}{z+1}\right]}\right),$$

if.
$\left[\frac{n+z}{z+1}\right] < \frac{n+z}{z+1}$ or $\left[\frac{n+z}{z+1}\right] = \frac{n}{z+1}$.
Case $\left[\frac{n+z}{z+1}\right] < \frac{n+z}{z+1}$ could occur, for example, if $n = 5, z = 2$.
Case $\left[\frac{n+z}{z+1}\right] = \frac{n}{z+1}$ could occur, for example, if $n = 6, z = 2$.

$$F_{s^*}(\delta) = C_{s^*}^{/1} \delta^{\frac{z(n+1)}{z+1}} + \vartheta\left(\delta^{\frac{z(n+1)}{z+1}}\right),$$

This case could occur, for example, if $n = 5, z = 2$.
Two cases should be considered.

(1) $\left[\frac{n+z}{z+1}\right] = \frac{n}{z+1}$.

(a) $\alpha_s z < \beta_s$, then $\alpha_s z < \left[\frac{n}{2}\right]$ where $\left[\frac{n}{2}\right]$ is the whole part of $\frac{n}{2}$ number and n is the number of navigation spacecrafts in the working constellation. By means of $F_s(\delta) - F_{s^*}(\delta) > 0$ transformations we come to a conclusion that, i.e., the offered working constellation with the account that it satisfies the restriction on the geometrical factor, really is the best according to the offered criterion.

(b) $\alpha_s z > \beta_s$, then $\beta_s < \left[\frac{n}{2}\right]$.

Let us consider the same ratio and come to the conclusion that $F_s(\delta) - F_{s^*}(\delta) > 0$, i.e., in this case, the proposed working constellation is also the best according to the proposed criterion.

(c) $\alpha_s z = \beta_s$, then $\alpha_s z \leq \left[\frac{n}{2}\right]$.

$$F_s(\delta) - F_{s^*}(\delta) > 0.$$

Thus, in the case of (1), only the proposed working constellation is the best according to the proposed criterion for the whole range of possibilities.

(2) $\left[\frac{n+z}{z+1}\right] = \frac{n}{z+1}$.

(a)   If $\alpha_s < \beta_s$, then $\alpha_s z < \left[\frac{n}{2}\right] < \frac{z(n+1)}{z+1}$.

Let us consider the difference

$$F_s(\delta) - F_{s^*}(\delta) = C_s^1 \delta^{\alpha_s z} - C_{s^*}^1 \delta^{\frac{z(n+1)}{z+1}}$$
$$- \vartheta\left(\delta^{\frac{z(n+1)}{z+1}}\right) + \vartheta\left(\delta^{\alpha_s z}\right)$$
$$= \delta^{\alpha_s z}\left(C_1 - \frac{\varphi(\delta^{\alpha_s z})}{\delta^{\alpha_s z}}\right),$$

where $C_1 = C_s^1$, $\varphi(\delta^{\alpha_s z}) = C_{s^*}^1 \delta^{\frac{z(n+1)}{z+1}} + \vartheta\left(\delta^{\frac{z(n+1)}{z+1}}\right) - \vartheta(\delta^{\alpha_s z})$.

By mathematical transformations with the use of the device of infinitely small functions we obtain

$$F_s(\delta) - F_{s^*}(\delta) > 0.$$

Therefore, the proposed working constellation is the best according to the proposed criterion.

(B) $\alpha_s z = \beta_s$, then $\alpha_s z \leq \left[\frac{n}{2}\right] < \frac{z(n+1)}{z+1}$,

where $C_1^{//} = C_s^{/1} + C_s^1$, $\varphi(\delta^{\alpha_s z}) = C_{s^*}^1 \delta^{z\frac{n+1}{z+1}} + \vartheta\left(\delta^{z\frac{n+1}{z+1}}\right) - \vartheta(\delta^{\alpha_s z})$.

Further, taking into account that $\varepsilon = \frac{C_1^{//}}{2}$, and, taking into account that $\varphi(\delta^{\alpha_s z})$ is the function infinitely small relative to $\delta^{\alpha_s z}$ we obtain the following:

$$C_1'' - \frac{\varphi(\delta^{\alpha_s z})}{\delta^{\alpha_s z}} > 0 \Rightarrow F_s(\delta) - F_{s*}(\delta) > 0.$$

Therefore, the proposed working constellation is the best according to the proposed criterion.

Thus, $\min_{s \in S} F_s(\delta) = F_{s*}(\delta)$ where $s*$ is a working constellation, which is a system of type $\left[\frac{n+z}{z+1}\right]$ out of n. Consequently, the above criterion shows that the working constellation, which is the system of type $\left[\frac{n+z}{z+1}\right]$ out of n is the best according to the proposed criterion.

In order to select the best working constellation, the user needs to have information about the angle of position of the navigation spacecraft, which can be provided by his or her receiving and indicating equipment and information about the accuracy of the characteristics, which can be taken from past statistical data of the use of the navigation spacecraft or navigation spacecraft data from various SRNS in the past. Once the onboard systems have been informed of all the navigation spacecraft that meet the above ratios, the pilot chooses the working constellation, which will be used to determine the aircraft coordinates as follows [15]:

(1) the navigation spacecraft that do not meet the geometric and accuracy requirements are discarded;
(2) the best four navigation spacecraft are selected among the remaining ones by the traditional geometric factor method.

The time factor to be taken into account is as follows: in the event of retransmission of the navigation information from the aircraft, the working constellation on which the previous information was transmitted may no longer be the best, as during this time the navigation spacecraft may go over the horizon or change the configuration of other working constellations [16, 17].

3.4 Shortcomings in the Integrity Monitoring Algorithm in the Receiver

RAIM used to calculate aircraft coordinates

Currently, the predominant use of satellite techniques onboard an aircraft is through the use of the receiver integrity monitoring algorithm (RAIM). RAIM makes several assumptions regarding the characteristics of the satellites it uses. The key ones are as follows: the possibility of erroneous data and complete absence of malfunctions in more than one satellite at the same time [10, 16].

A recent analysis of the GLONASS characteristics has shown that the probability of malfunctions on GLONASS satellites can be quite high, as well as the presence of events where several satellites had large errors at the same time. The RAIM algorithms currently in use would not be able to meet the required levels of integrity if they processed GLONASS satellite data in the same way that they processed GPS satellite data [18–20].

RAIM modifications are available to allow different assumptions to be applied to GPS and GLONASS. The proposed changes allow the user to benefit from this second constellation while maintaining the required level of integrity. In particular, the possibility of several failures in the constellation GLONASS is taken into account. Horizontal guidance is the service that RAIM provides today. However, we recommend that the onboard equipment algorithm can provide assurance that the horizontal positioning error (HPE) is less than 0.3 m. mile within the required integrity level. This is the value that RAIM currently provides for flights from the route to the airfield phase and is currently provided by RAIM.

The Global Positioning System Standard Location Service specifications provided assurance that there would be no more than three major operational failures per year. In 2008, the Characteristics Standard of the Standard Location Service of the Global Positioning System was updated and redefined, with a significant operational failure being a signal in space error (SIS) of 4.42 times the transmittable accuracy of the range measurement (URA) of 2.4 m.

The onboard equipment RAIM algorithm uses measurement redundancy to identify SIS errors. The measurement data for each satellite are compared with aggregate information from other satellites in turn. Should data inconsistencies be found, the algorithm will try to link this error to a specific satellite and then remove this satellite from the sample. If the algorithm is unable to detect this satellite, an error message will be displayed. The algorithm also evaluates the largest possible non-detectable error and limits it to a value called the horizontal protection level (HPL). The higher is the SIS accuracy and the higher is the measurement redundancy, the lower is the HPL value. Specific flight stages involve corresponding maximum permissible errors called horizontal alarm limits (HAL). If the HPL is lower than the HAL, the flight is assumed possible [13, 21].

In order to add GLONASS to the onboard RAIM algorithms, it is essential to understand how GLONASS works in comparison with GPS and the assumptions of the algorithm. The accuracy and reliability of SIS are two very important characteristics.

GPS satellites have relatively smaller errors. SIS GLONASS errors have statistically high mean values and a larger spread. Average GPS errors are usually a few centimeters or less, while average GLONASS errors can be more than a meter. Standard GPS deviations are close to half a meter for recently launched satellites. The range of standard deviations of GLONASS is from less than one meter to almost two meters. On average, GLONASS errors are two or three times greater than GPS errors. This is partially reflected in the transmitted URA values. The most common URA GPS value is 2.4 m (the lowest possible value), while the most common URA GLONASS value is 4 m.

The malfunction frequency for GLONASS is at least an order of magnitude higher than the malfunction frequency for GPS. It is also obvious that GLONASS is improving and the overall malfunction rate is decreasing. It is possible that in a few years the malfunction rates of these two constellations will be comparable. It should be assumed beforehand that the probability that the GPS is in a malfunctioning state is 10^{-5} (an approximate value used by the existing RAIM algorithms), and for the corresponding probability for GLONASS we will use the value 10^{-4}.

The ARAIM concept is based on the fact that in the future multistellar (multiple satellite systems) and multifrequency signals will decrease the dependence on the ground infrastructure and, as a result, will further reduce the implementation and operation costs. The user will be provided with an augmentation system signal (ISM integrity support message), which is updated and transmitted with a long waiting time (delay). This is the main idea of advanced autonomous control in the receiver [22–24].

The aim of the ARAIM concept is to provide aviation users with the ability to orientate themselves vertically up to an accurate approach based on multistellar Global Navigation Satellite System (GNSS) signals. Navigation systems that support the possibility of vertical guidance of aircraft are subject to several requirements that determine their characteristics.

These requirements are presented below:

(1) Reliable vertical accuracy of 4 m at 95% and 10 m at 10^{-7},
(2) Vertical accuracy with 15 m failure probability at 10^{-5},
(3) Vertical error limit of 15 m with a risk of integrity loss of 0.5 (vertical alert, VAL),
(4) Risk of damage to the integrity equal to $2 \cdot 10^{-7}$ of the landing approach (150 s),
(5) The trigger alert time (TTA) is 6 s.

The purpose of a terrestrial system in ARAIM is to provide integrity parameters for users known as integrity support message (ISM). In comparison to the classical RAIM, which represented a constant assumption on the characteristics of signals in space (SIS), the ARAIM architecture is more flexible: a special ground-based system monitors the characteristics of GNSS and adapts the integrity parameters.

The ARAIM algorithm works by evaluating different malfunction modes with predetermined malfunction probabilities and determining the optimal probability of non-detection for each mode (some modes may already be quite unlikely and do not require special testing). For those of them that do require an assessment, an appropriate subset of satellites is selected and the resulting location solution is compared to the "all visible satellites" solution. A subset of satellites is formed, and all satellites that have malfunctions as a result of the assessment are excluded. Thus, a subset is created to be free from malfunctions and, being compared with the "all visible satellites" solution, detects a failure of the integrity of the suspected malfunction. Provided that the results of all comparisons are satisfactory, the HPL is calculated, as well as the location "all visible satellites" solution. If one of the scores is unsatisfactory, the algorithm needs to make an exception to remove the defective satellites or declare that a safe solution is not possible [14, 22].

In developing ARAIM, the aim was to carefully consider all the threats that are likely to be identified. As mentioned earlier, the RAIM option used today takes into account only one rather probable threat: a malfunction of one satellite leading to a range measurement error greater than the calculated one. A case where one reason causes great errors in distance measurement on more than one GPS satellite is considered to be quite unlikely. For GLONASS, both a malfunction affecting the operation of one satellite and a malfunction affecting several satellites are considered. Like traditional RAIM, special malfunction modes are considered for ionospheric, tropospheric, or local multi-path effects. It is assumed that all of them create horizontal positioning errors, which are significantly lower than the horizontal alarm threshold of 556 m.

Table 3.6 shows the malfunction modes.

The ARAIM algorithm evaluates the subsets corresponding to each mode. Each check will have a reasonable chance of being undetected. Safety is ensured if the product of the non-detection value and a priori probabilities is less than the required value [23].

The first malfunction mode (Mode 1) should be assessed in the same way as currently performed in RAIM. Due to the fact that the threat of malfunction on a constellation-wide GPS scale is not considered, Modes 3, 8, 10, and 12 have a zero a priori probability of occurrence and do not need further assessment. The malfunction on the constellation scale of GLONASS is assessed by comparing the location solution with a location solution obtained only using GPS satellites.

This creates a subset that is not affected by any of the GLONASS malfunction modes. This test evaluates not only the entire constellation-wide malfunction mode,

Table 3.6 Malfunction modes

Mode	Description
0	No malfunction
1	Malfunction of single SV GPS
2	Malfunction of single SV GLONASS
3	Malfunction of multiple SV GPS
4	Malfunction of multiple SV GLONASS
5	Two independent single malfunctions of SV GPS
6	Two independent single malfunctions of SV GLONASS
7	Malfunction of single SV GPS and Malfunction of single SV GLONASS
8	Malfunction of single SV GPS and malfunction of multiple SV GPS
9	Malfunction of single SV GPS and malfunction of multiple SV GLONASS
10	Malfunction of single SV GLONASS and malfunction of multiple SV GPS
11	Malfunction of single SV GLONASS and malfunction of multiple SV GLONASS
12	Malfunction affecting several SV GPS and malfunction affecting several SV GLONASS
13	Three or more partially overlapping independent malfunctions

but also any malfunction mode that affects any number of GLONASS satellites. Consequently, Modes 2, 4, 6, 11 and any associated higher order fault modes affecting only GLONASS are included in this assessment. Therefore, the a priori probability of this malfunction mode is the total number of these modes [17].

The remaining Modes 5, 7, and 9 may not need assessment provided that there are not too many satellites.

It is recommended that the ARAIM algorithm assess the trouble-free mode (all satellites included in the location calculation) with an a priori probability margin of one. It also evaluates the subsets from which each individual GPS satellite is removed, with an a priori probability margin of 10^{-5}.

If there are no visible GLONASS satellites, there is no need to evaluate this additional mode, and the default algorithm returns to the mode that corresponds to modern algorithms. If there are not enough visible GPS satellites (N GPS < 4) to calculate the location, the algorithm cannot work because the high probability of a GLONASS malfunction covering the entire constellation requires prevention of consequences by comparing the GPS-only location solution with the "all visible satellites" solution.

References

1. Bakulev PA, Sosnovskiy AA (2011) Radio navigation systems. Radiotehnika
2. Didenko NI, Eliseyev BP, Sauta OI, Shatrakov AY, Yushkov AV (2017) Radio-technical support of civil and military aviation flights: strategic problem of the arctic region of Russia. Nauchniy Vestnik of MSTUCA 20(5):8–19
3. Zatuchnyy DA (2008) Build-up method for an optimized working constellation based on relia-bility criteria with regard to two methods of integrity control. In: Proceedings of the international symposium "Nadyozhnost i kachestvo", Penza, vol 1, pp 307–309
4. Zatuchnyy DA (2007) Building-up of an optimized SRNS based on reliability criteria - Nauchniy Vestnik of MSTUCA, series "Radiofizika i radiotekhnika". No. 112, pp 106–108
5. Zatuchnyy DA (2010) Optimized controller-pilot data link communications system build-up. In: Proceedings of the international symposium "Nadyozhnost i kachestvo". Penza, vol 1, pp 430–431.
6. Zatuchnyi DA (2018) Analysis of various jamming impact on the civil aircraft navigation systems. Informatizaciya I Svyaz. 2:7–11
7. Zatuchnyy DA, Logvin AI, Nechayev EE (2012) Problems of ADS mode implementation in the Russian Federation. Printing and publication department MSTU CA
8. Karyukin GE (2005) Improvement of the navigation sighting methods in satellite radio navi-gation systems when solving civil aviation navigation problems. Interuniversity collection of scientific works "Problemy ekspluatacii i sovershenstvovaniya transportnyh sistem", vol. XI. St. Petersburg: Academy of CA
9. Kozlov AI, Garanin SA (2005) On the development of mathematical simulations reflecting impact of radio jamming and random effects affecting the aircraft on the navigation parameters to be determined. Nauchniy Vestnik of MSTUCA, series Radiofizika i radiotekhnika, no. 93.
10. Kuzmin BI, Meshalov RO (200) Implementation of CNS/ATM systems in the RF CA. Magazine "Elektrosvyaz", no. 5
11. Logvin AI, Orlov OE (2002) Satellite radio navigation and communication systems for ATC system. M.: MSTUCA
12. Nechayev EE, Budykin YA (2005) Aerial devices in civil aviation. Kursk, "PRESS-FAKT"

13. Manual on airspace planning methodology for the determination of separation minima (1998) ICAO, First Revision
14. Required navigation performance (RNP) Manual (2008) Third Revision
15. Sizyh VV, Shahtarin BI, Shevcev VA (2017) Cycle slip mechanism in stochastic analog systems of first- and second-orders of phase-lock. Mekhatronika, avtomatizaciya, upravlenie, vol 18, no. 1
16. Solovyov YA (2000) Satellite radio navigation system. Moscow
17. Solomencev VV, Garanin SA (2003) Ways to improve reliability of AK and AS ATC software. Nauchniy Vestnik of MSTUCA, series Informatika. Prikladnaya matematika, no. 65
18. Solomencev VV, Fedotova TN, Garanin SA, Ignatenko OA (2003) ATC automation systems complexes reliability. Nauchniy Vestnik of MSTUCA, series Informatika. Prikladnaya matematika, no. 65
19. Tikhonov VI (1982) Statistical radio engineering. M.: Radio i Svyaz, p 624
20. Tikhonov VI, Kulman NK (1975) Non-linear filtering and quasi-coherent signal reception. M.: "Sov. Radio", p 704
21. Tikhonov VI, Mironov MA (1977) Markoff processes. M.: Sov. Radio, p 488
22. Tikhonov VI, Kharisov VN (199) Statistical analysis and synthesis of radiotechnical devices and systems. M.: Radio i Svyaz, p 608
23. Falkov E, Shavrin S (2017) Cyber security of aircraft information and communication systems. Informational and analytical journal "Radioelektronnye tekhnologii", no. 5
24. Filin AD, Shatrakov YG, Yakovlev VT, Yushkov AV (2017) The impact of flight support subsystems reliability on probability of aircraft accidents. Nauchniy Vestnik of Russian military-industrial complex, no. 3, pp. 68–74.

Chapter 4
Quality Enhancement of Data Transmission via Civil Aircraft Communication Systems by Proper Use of Communication Resources

4.1 Principle of Guaranteed Sufficiency of Communication System Using

In accordance with the principle of guaranteed sufficiency, a rational demand for communication resources to support aircraft flights shall be determined. This demand is determined, on the one hand, by the intensity of air traffic and the jamming environment, and, on the other hand, by the intensity of data transmission determined by the accepted service procedure and the current flight environment.

The principle of guaranteed sufficiency presupposes a qualitative assessment of the required communication resources to support aircraft flights under certain conditions. However, the probabilistic nature of these parameters does not allow unambiguous determination of the required resources [1].

As the principle of guaranteed sufficiency suggests, quantitatively determined resources should provide requirements for ensuring flight safety with a given probability. This condition can be defined by the following formula [2]:

$$P(d = 0) \geq \gamma, \qquad (4.1)$$

where γ is a specified probability of ensuring compliance with the safety requirements, and d is the number of service failures.

The intensity of air traffic is related to the message transmission rate by the non-linear functional dependence. The non-linear nature of this dependence is due, on the one hand, to the finite bandwidth of the communication channel, and, on the other hand, to the limited psychophysical capabilities of a traffic controller. The simplest convenient mathematical model that allows considering the non-linear nature of the functional dependence of the "upper bound" type is the exponential dependence [3]

$$Y = Y_m * \left[1 - e^{-x/x_m}\right]. \qquad (4.2)$$

© The Author(s), under exclusive license to Springer Nature Singapore Pte Ltd. 2021
D. A. Zatuchny et al., *Noise Resistance Enhancement in Aircraft Navigation and Connected Systems*, Springer Aerospace Technology,
https://doi.org/10.1007/978-981-16-0630-4_4

There are indications for the following: Y_m is the maximum possible message transmission rate for this type of communication channel, x_m is the highest possible intensity of air traffic providing no overload of the communication channel. Considering the relation that follows from (4.2)

$$\frac{dY}{dX}\big|_{x=0} = \frac{Y_m}{X_m},$$ (4.3)

X_m can be determined through Y_m, and the specific message transmission rate per unit of measurement of the air traffic intensity when there is no data channel overload

$$V_0 = \frac{dY}{dX}\big|_{x=0}.$$ (4.4)

Additionally, considering (4.3) and (4.4), we have

$$X_m = \frac{Y_m}{V_0}.$$ (4.5)

The parameters Y_m and V_0 are sufficiently physical. In particular, Y_m depends on the communication channel bandwidth and psychophysical capabilities of a traffic controller. As for the parameter V_0, it is determined by the operating conditions and may vary depending on the flight conditions (normal conditions, a thunderstorm bypass, etc.), for example. Factually, V_0 determines the amount of data to be transmitted via data transmission lines from a single aircraft for the given operating conditions. With respect to the above, the parameter V_0 may be determined as follows:

$$V_0 = \frac{q_1}{y_{cp}},$$ (4.6)

where q_1 is the amount of data to be received from the aircraft within the air traffic control (ATC) zone, and y_{cp} is the average time of aircraft stay within the zone.

Thus, considering (4.5), we have

$$X_m = Y_m * \frac{y_{cp}}{q_1}.$$ (4.7)

Functional communication characteristics are usually calculated for centered random processes. In this case, they are calculated for deviations in air traffic intensity and message transmission rate from the average values of X_0 and Y_0 as

$$\begin{cases} \xi = X - X_0 \\ \nu = Y - Y_0 \end{cases}.$$ (4.8)

At the same time, if we take ξ and v as the parameters for which the functional communication characteristics are calculated, considering (4.2) and (4.8), the direct and inverse functional transformations may be represented as

$$v = Y_m * \exp\{-X_0/X_m\}\left[1 - \exp\{-\xi/X_m\}\right] \tag{4.9}$$

$$\xi = -X_m \ln\left[1 - (v/Y_m) * \exp\{X_0/X_m\}\right] \tag{4.10}$$

or, considering (4.7), as

$$v = f(\xi) = Y_m * \exp\{-X_0/Y_m y_{cp}\}\left[1 - \exp\{-\xi * q_1/Y_m y_{cp}\}\right], \tag{4.11}$$

$$\xi = \varphi(v) = [Y_m y_{cp}/q_1] * \ln[1 - (v/Y_m) * \exp\{X_0 q_1/Y_m y_{cp}\}]. \tag{4.12}$$

Figure 4.1 shows the dependence between the radio exchange and the air traffic intensity.

The mathematical model of functional communication can be significantly simplified if we take the standardized maximum possible intensity of air traffic as the parameters when there is no data channel overload (air traffic intensity is measured from the average value) [4]:

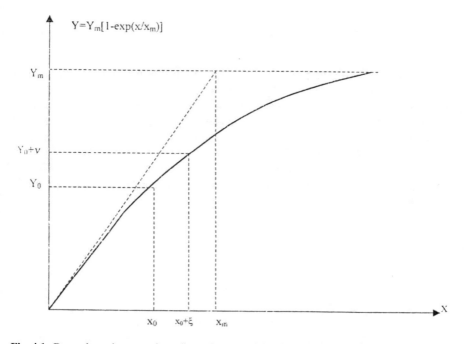

Fig. 4.1 Dependence between the radio exchange and the air traffic intensity

$$X = \xi / X_m = (X - X_0)/X_m = (X - X_0)q_1/Y_m y_{cp}, \qquad (4.13)$$

and the ratio of the average messaging reserve to the actual.

$$\mu = (Y_m - Y_0)/(Y_m - Y_0 - \nu) = (Y_m - Y_0)/(Y_m - Y). \qquad (4.14)$$

In this case, the direct and inverse transformations that connects X and μ are determined by the expression

$$\mu = f(X) = e^x, \qquad (4.15)$$

$$X = \varphi(\mu) = \ln \mu. \qquad (4.16)$$

The situations considered during flight are classified into four groups according to the coordinate data update rate:

(1)   data update rate for the first group is 5 c;
(2)   data update rate for the second group is 7 c;
(3)   data update rate for the third group is 30 c;
(4)   data update rate for the fourth group is 180 c.

The number of channels required to comply with the requirements for flight safety can be determined by the formula [5]:

$$l = f(n, \alpha, \tau), \qquad (4.17)$$

where n is the number of aircraft within the control area of the traffic controller, α is the aircraft flight situation ($\alpha = 1, \ldots, 4$), and τ is the period required to reassign the communication channel from one aircraft to another one. We assume that the redundant communication channels are loaded, i.e. the communication channel is reassigned from the aircraft not transmitting the message to the aircraft, which shall transmit the message. Assume that $\tau = $ const.

Now let's formulate a criterion for determining the number of channels required to comply with the requirements for flight safety, in accordance with the principle of guaranteed sufficiency:

$$l \rightarrow \min \qquad (4.18)$$

with the following limitation:

$$P(d = 0) \geq \gamma.$$

One of the components of the number of communication channels that are permanently assigned to the aircraft being in 1–3 situations is a value calculated by the formula:

$$n_1 = n(p_1 + p_2 + p_3), \tag{4.19}$$

where p_1, p_2, p_3 are the probabilities of finding the aircraft in the first, second, and third situations, respectively, and n is the number of aircraft under the control of the traffic controller.

Let us introduce the following symbols:

$p_i^/$ is a probability of malfunction-free operation i of channels assigned to i aircraft, $i = 0, \ldots, n_1$;

$q_i^/$ is a probability of malfunction i of channels assigned to i aircraft, $i = 0, \ldots, n_1$.

Then the mathematical expectation of the number of malfunction-free communication channels is calculated by the formula:

$$m = \sum_{i=0}^{n_1} i p_i^/. \tag{4.20}$$

We conclude that $m = n_1$, since the probability of malfunction-free operation of the communication channel is a value that can be considered approximately equal to one.

Let us introduce the following value:

$$m_1 = \max i : q_i^/ > 1 - \gamma. \tag{4.21}$$

Accordingly, the required number of redundant channels required to comply with the flight safety requirements equals m_1.

The event consisting in the failure-free servicing of the requirements comprises one of two events: malfunction-free operation of the communication channel or failure of the communication channel and malfunction-free switching to another channel:

$$P(d = 0) = P_{\text{op.}} + P_{\text{malf.}} \sum_{i=0}^{k-1} (1 - P_{\text{switch.}})^i P_{\text{switch.}}, \tag{4.22}$$

where

P_{op} is a probability of communication channel malfunction-free operation;

$P_{\text{malf.}} = 1 - P_{\text{op.}}$ is a probability of communication channel malfunction;

k is the maximum number of communication channels available;

$P_{\text{switch.}}$ is a probability of malfunction-free switching to another channel.

The number of terms within expression (4.22) is determined by expression (4.1). Let us introduce the following symbols:

$$\sum_{i=0}^{k-1} (1 - P_{\text{switch.}})^i P_{\text{switch.}} = Q,$$

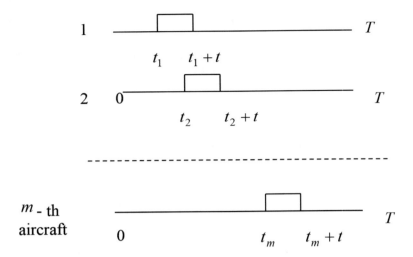

Fig. 4.2 Distribution of messages from the aircraft during the period of a traffic controller operation

$$q = \min i : P_{op.} + Q \geq \gamma.$$

For aircraft in the first, second, and third situations (the communication channel is assigned to the aircraft permanently), the number of communication channels can be determined by the formula:

$$l_1 = n_1 + m_1 \cdot q. \tag{4.23}$$

For the aircraft in the fourth situation, we introduce the following designations:
T is a period of a traffic controller operation;
t is a message duration.

Figure 4.2 shows the distribution of messages from the aircraft in the fourth situation during the period of a traffic controller operation.

Let us introduce the following symbols:

$$T_1 = [t_1; t_1 + t], T_2 = [t_2; t_2 + t], \ldots, T_m = [t_m; t_m + t].$$

There is a set of conditions under which, for a given aircraft 1 flight situation, the communication channel serves m aircraft:

$$\begin{cases} \overline{\dfrac{T_1 \cap T_2}{T_3 \cap (T_1 \cup T_2)}} \\ \ldots\ldots\ldots\ldots\ldots\ldots\ldots \\ T_m \cap \left(\bigcup T_i\right), i = 1, \ldots, m - 1 \end{cases}. \tag{4.24}$$

Let us introduce the following symbols: $g_1 = P\left(\overline{T_1 \cap T_2}\right) = 1 - \frac{t}{T}$, $g_2 = P(T_3 \notin T_1 \cup T_2), \ldots, g_{m-1} = P(T_m \notin \cup T_i, i = 1, \ldots, m - 1)$, $G = g_1 g_2 \ldots g_{m-1}$.

Then, if $G \geq \gamma$, then according to the principle of guaranteed sufficiency, taking into account (4.1), channel 1 can serve m aircraft in the fourth situation. As a rule, taking into account the frequency of messages, their average duration, and the traffic controller operation period, this situation may be considered impossible; thus, more than one communication channel has to be available. The following rule may be used to determine the number of channels:

(1)   Calculate the probabilities of all possible combinations of intersections per 2 messages, 3 messages, etc.;
(2)   Select the situation with the maximum probability;
(3)   The number of redundant channels shall correspond to the number of intersections for the situation with the maximum probability.

Let us introduce the following symbol:
m_2 is the number of redundant channels.

Now we determine the number of channels required to comply with the safety requirements of flights, in accordance with the principle of guaranteed sufficiency for the aircraft in the fourth situation provided that one communication channel serves all aircraft:

$$l_2 = 1 + m_2 q. \tag{4.25}$$

Then the total number of channels required to ensure the principle of guaranteed sufficiency is calculated by the formula:

$$l = l_1 + l_2. \tag{4.26}$$

There is an example showing the calculation of the required number of channels to ensure the principle of guaranteed sufficiency using the example of the Moscow region. There are the required initial data:

(1)   traffic controller operation period 'τ';
(2)   average message transmission period $t = 5c$;
(3)   probability of communication channel malfunction-free operation 0.999;
(4)   probability of malfunction-free switching to another channel 0.999;
(5) $\gamma = 0.999999$;
(6)   number of aircraft under the control of the traffic controller $n = 11$.

The number of aircraft in the first three situations equals 3, and the number of aircraft in the fourth situation equals 8.

Calculate the terms included in expression (4.22):
1 term equals 0.999 and according to the formula (4.22):

$$P(P_{\text{omk}} = 0) = 0.999 + 0.001 \cdot 0.999 = 0.999999.$$

Thus $q = 1$.

Calculate the value m_1 according to the formula (4.23):

$q_0' = 0.999^3 = 0.997, q_1' = 3 \cdot 0.999^2 \cdot 0.001 = 0.003, q_2' = 3 \cdot 0.999 \cdot 0.001^2 = 0.000003$,

$q_3' = 0.001^3 = 0.000000001$.

As only $q_3' < 1 - \gamma$, then $m_1 = 2$.

Thus,

$$l_1 = 3 + 2 \cdot 1 = 5.$$

Considering that the frequency of message transmission for the fourth situation equals $180c$ and the number of aircraft in the fourth situation equals 8, then the number of messages from all aircraft for the period of the traffic controller operation is calculated as follows:

$$4ч = 240\,мин = 14400c,$$

$14400 : 180 = 80$ is the number of messages transmitted from one aircraft;

$80 \cdot 8 = 640$ is the number of messages transmitted from all aircraft.

As $g_1 = 1 - \frac{t}{T} = 1 - \frac{5}{14400} < \gamma$, 1 communication channel to service all aircraft is insufficient.

Among all intersection probabilities, the situation with two intersections has the biggest value 0.84. Then

$$m_2 = 1,$$

$$l_2 = 1 + 1 \cdot 1 = 2,$$

$$l = 5 + 2 = 7.$$

The program for selecting the data transmission line is given in Appendix B.

4.2 The Principle of "Zonal" Use of Communication Systems

The principle of "zonal" use of civil aircraft flight support communication systems is based, on the one hand, on the concept of joint resources, and, on the other hand, on the basis of consideration of all features of the airspace of a given zone in order to ensure the rational use of these resources in the relevant zone [6].

The concept of joint resources is built on the basis of integration of all resources in a given zone and the resources of all communication systems in order to fulfill the general task of aircraft flight support. In this case, it is supposed to integrate all the resources involved in processing and transmitting data, in particular, communications and computing resources, within each system.

The concept of joint resources leads to a strategy for the integrated use of different communication systems in combination with each other, as well as their integrated use with navigation and surveillance systems. The features of each airspace zone shall include an individual set of ground-based technical equipment. Tactical performance of communication equipment, for example, range of action, initially determined only by technical parameters, largely depends on the quality of the equipment itself [3, 7].

Currently, there is an objective necessity to manage joint resources. At the same time, the aircraft flight support system for the management of these resources shall be an integral subsystem of a large technical system, which is the system of air traffic control. Such systems have the following specific features: complexity of behavior, hierarchy of behavior, and multi-purpose nature of external impacts.

Based on the foregoing, it may be affirmed that the flight support management system of joint resources shall have a hierarchical structure with a multi-purpose control character. At least two objectives of such management can be distinguished, which shall define, respectively, two stages of management. The first stage is a system structural management stage, which determines the number of communication channels used. The second stage is a channel control stage, which manages the level of the assigned channels' occupancy ensuring the transmission of both basic and non-basic data.

In the aircraft flight support management system of joint resources, change in the air traffic intensity serves as the main external influence, which actually determines the whole system operation algorithm.

The probabilistic nature of this change is a major obstacle to solving the problem of managing joint resources. In this case, one of the main issues is to forecast load changes within the communication system. Since the data load at aircraft flight support is constantly changing, and not only due to changes in the intensity of air traffic, but also due to changes in flight conditions (changes in weather conditions, aircraft approach, etc.), the control system of joint resources shall be a system for information collection and transmission. One of the main issues in the system for information collection and transmission is the accuracy of forecasting the required communication resources and selection of service procedures. A priority service procedure which provides servicing of top-priority sources takes a special place.

Thus, such a principle of managing communication resources implies an automatic redistribution of communication channels' number and their bandwidth depending on the volume of data load, activity of data sources, and data load value.

The principle of zonal use of communication systems at aircraft flight support involves forecasting the message transmission rate within the zone and assessing the influence of jamming situation within the zone on the intensity of communication systems' use.

The data exchange rate λ'_p in the ith zone depends on the intensity of air traffic $X(t)_i$ in this zone and its ground support characterized by the equipment status factor K_i, i.e. [8].

$$\lambda'_p = f[X(t)_i, K_i], \tag{4.27}$$

where f is a certain function. For the forecasting of λ'_p, it is necessary to predict the intensity of air traffic in this zone.

The resulting forecast λ'_p is further multiplied by the corresponding coefficient K_i characterizing the necessary increase in the transmission of messages in the ith zone depending on its equipment status. According to the obtained message transmission rate, the necessary joint resources are determined.

It is reasonable to trace the jamming environment's influence on the operating performance problem when determining the accuracy of the aircraft position tracking.

The system operation quality can be characterized by the probability P that the error in the aircraft position tracking shall not be exceeded at ADS Δx_{A3H} of permissible error ' Δx_{don} ' with the preset value ' P_{zad} ', e.g. [5, 9]

$$P\left(\Delta x \leq \Delta x_{доп}\right) \geq P_{зад.} \tag{4.28}$$

The accuracy of the aircraft location tracking by any navigation equipment is determined by the measurement error Δx_u. The measurement error depends on the accuracy of the navigation equipment and the nature of the navigation data processing using the onboard navigation instrumentation (ONI). This error is random and is included in the summary error of the aircraft position tracking.

The second group of factors affecting the error Δx is associated with the transmittance discreteness of data on the aircraft position. According to the transmitted discrete messages, the continuous trajectory of the aircraft movement is restored by extrapolation in the ATM center.

In real-world conditions, the aircraft can perform a controlled maneuver and be exposed to atmospheric disturbances. The error in position tracking ' Δx_∂ ' arising from this shall be called dynamic since it is determined by the dynamics of the aircraft motion. In this case, there shall be a maximum dynamic error ' Δx_∂ ' in case of an assumption that the aircraft carries out the maximum allowable maneuver under the most adverse atmospheric effects immediately after the transmission of the message. When the civil aircraft is moving along the route, the only allowable maneuver, apart from altitude level change, is the horizontal orbit, which is carried out by the aircraft's circular motion by changing the roll angle. In this case, the maximum intensity of the orbit is determined by the minimum turning radius R_0 determined, in turn, by the maximum allowable roll angle γ and the maximum true airspeed V_u:

$$R_0 = \left(\frac{V_u^2}{g}\right) tg\, \gamma, \tag{4.29}$$

where $g = 9.8\ \text{м}/c^2$ is the gravity acceleration.

The most unfavorable atmospheric impact in respect of the effect on the aircraft lateral divergence from the assigned track (AT) is the wind directed along the normal to the extrapolation line with the greatest speed. The maximum lateral divergence of

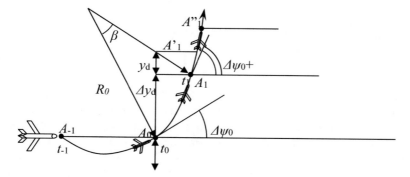

Fig. 4.3 Aircraft maneuvering

the aircraft from the extrapolation line within the period of Δt is determined by the expression

$$\Delta x_{\partial}(\Delta t) = R_0[\cos\Delta\psi_0 - \cos(\Delta\psi_0 + \beta)] + U_{\max}\Delta t \tag{4.30}$$

where $\Delta\psi_0$ is the maximum possible deviation angle of the aircraft from the extrapolation line during the transmission of the previous message, β is the maximum angle of rotation of the aircraft within the period Δt determined by the expression

$$\beta = V_u \cdot \Delta t / R_0, \tag{4.31}$$

$U_{\max}$ is the maximum wind speed directed along the normal to the extrapolation line, Δt is the period from the moment of the message transmission start to the moment of the next message reception, and the other designations are the same. An example of the aircraft maneuvering is shown in Fig. 4.3.

Another group of factors affecting the Δx is associated with a delay in the message transmission via communication channels τ_3 resulting from a delay in the signal propagation τ_p, delays in processing the message by the equipment τ_{an}, delays in the analysis $\tau_{a\mathcal{H}}$ determined by the time required by the ATM center for data comprehension, and delays when sending a message via the communication network τ_c associated with the message re-transmission when errors are detected, queueing delay at network overload, message switching when sending them over traffic that contains several heterogeneous communication channels.

As a result of the delay during message transmission, the extrapolation time shall increase and equal

$$\Delta t = T + \tau_3, \tag{4.32}$$

where T is the update period,

$$\tau_3 = \tau_p + \tau_{an} + \tau_{a\mathcal{H}} + \tau_c. \tag{4.33}$$

Accordingly, the dynamic position tracking error of the aircraft shall increase, which shall lead to an increase in the summary error:

$$\Delta x = \Delta x_{\text{и}} + \Delta x_{\text{д}}. \tag{4.34}$$

The most uncertain is the delay when transmitting a message via communication network τ_c since it depends on many factors. So, the Bose–Chaudhuri–Hocquenghem (BCH) code commonly used in data transmission channels, during error detection in the message, does not allow correcting them. In this case, it becomes necessary to re-transmit the message, which prolongs the delay for the data update period T, with automatic transmission of messages within approximately 10 s. The probability of such a random delay is determined by the probability of the message transmission malfunction and depends on the noise immunity of the communication channel used [10].

Let us make the following remark: in order to reduce the total error during ADS, it is necessary to reduce two types of errors:

(1)   error caused by the navigation equipment error Δx_u;
(2)   dynamic aircraft position tracking error ' $\Delta x_{\partial}(\Delta t)$'.

Since it is assumed that the wind speed U_{max} is a value that does not depend on the actions of the aircraft crew, then the dynamic error can be reduced in three ways:

(1)   reduce the minimum turning radius R_0;
(2)   reduce the value $\cos \Delta \psi_0 - \cos(\Delta \psi_0 + \beta)$;
(3)   reduce the period from the moment the message is sent

to the moment the next message is received.

To reduce the value R_0, it is necessary to reduce the aircraft true speed and roll angle; to reduce the value $\cos \Delta \psi_0 - \cos(\Delta \psi_0 + \beta)$, it is necessary to reduce the angle d_i. In this case, the greatest contribution to the dynamic error of the aircraft is made by the true speed of the aircraft, then an increase in the roll angle, and then an increase in the message transmission period.

It is desirable to reduce the value,

R_0 in the first place, as it may increase without limit, and $\cos \Delta \psi_0 - \cos(\Delta \psi_0 + \beta) \leq 2$.

Let us introduce the following symbols:

$$G(V_u, \gamma) = R_0 = \frac{V_u^2}{g} tg\gamma,$$

$$F(\Delta \psi_0, \beta) = \cos \Delta \psi_0 - \cos(\Delta \psi_0 + \beta).$$

After making the necessary transformations, we get

$$F(\Delta \psi_0, \beta) = 2 \sin \frac{\beta}{2} \sin \left(\Delta \psi_0 + \frac{\beta}{2} \right). \tag{4.35}$$

Obviously, this value shall be minimal if $\beta \to 0$. This condition coincides with the condition of reducing the minimum turning radius, as well as reducing the time interval for message transmission. In case if $\Delta\psi_0 \approx 0$, then $F(\Delta\psi_0, \beta) = \frac{\beta}{2}\left(\Delta\psi_0 + \frac{\beta}{2}\right)$, and if the ratio limit of $\Delta\psi_0$ to β is zero, then $F(\Delta\psi_0, \beta) \approx \frac{\beta^2}{4}$. If we take other variables V_u and γ, by making a replacement $F(\Delta\psi_0, \beta) = F(V_u, \gamma) = \frac{\Delta t^2 g^2}{4V_u^2 t g^2 \gamma}$, then we can formulate a criterion, following which the aircraft crew can reduce the dynamic error:

$$\min G(V_u, \gamma)$$

$$F(V_u, \gamma) \le \alpha,$$

where α is a certain number.

The following recommendation follows from this criterion: in order to reduce the dynamic error, it is necessary to reduce the true speed and the roll angle of the aircraft to the acceptably possible level that satisfies the boundary condition.

If we suppose that the true speed of the aircraft is a relatively constant value, then the value R_0 may be represented as $R_0 = C_1 t g \gamma$, and given that it is desirable to minimize the roll angle, then $tg\gamma \approx \gamma$ and $R_0 = C_1 \gamma$. In this case, the value β can be represented as $\beta = C_2 \frac{\Delta t}{\gamma}$. Then the criterion for reducing the dynamic error can be represented as

$$\gamma \to 0,$$

$$\frac{\Delta t}{\gamma} \le \alpha.$$

Thus, we draw the following conclusions [11–14]:

(1) In general, the dynamic error depends on the true speed of the aircraft, the roll angle of the aircraft, and the period from the moment of the ADS message transmission start to the moment of the next message reception.
(2) The dynamic error of ADS is directly proportional to the roll angle of the aircraft and has a quadratic dependence on the true speed of the aircraft.
(3) To reduce the dynamic error at an identified true speed of the aircraft, it is necessary to reduce the roll angle of the aircraft to a level that satisfies the restriction of the ratio between the period from the moment of the message transmission start to the moment of the next message reception and the roll angle.

References

1. Aeronautical Telecommunications Annex 10 to the Convention on International Civil Aviation—Montreal, ICAO, sixth revision, July 2006
2. Zatuchnyy DA (2010) Optimized controller-pilot data link communications system build-up. In: Proceedings of the international symposium "Nadyozhnost i kachestvo", Penza, vol 1, pp 430–431
3. Akinshin RN, Menshikov VL, Sushkov AV, Shamanayev AV, Pogrebezhskiy AG (2006) Evaluation of jam resistance of wideband radio communication systems during internal and external jamming. In: XXIV-th scientific conference dedicated to the Radio Day (collection of scientific works), Tula, pp 82–86
4. Akinshin NS, Tarkhov NS (2007) Hardware and software complex for research of aerial systems and communication channel via orbital artificial satellite. In: XV-th international conference proceedings "Radiolokaciya i radiosvyaz". Institute of Radio Engineering and Electronics RAS, Moscow, pp 502–506
5. Zatuchnyy DA, Logvin AI (2013) Determination of data links optimal number for ADS mode implementation. Nauchniy Vestnik of MSTUCA 189:9–13
6. Akinshin RN, Andreyev AV, Rumyancev VL, Esikov OV (2016) Application of a genetic algorithm for selection of air traffic control system radio-technical facilities operating frequencies, vol 19, no 5. Nauchniy Vestnik of Moscow State Technical University of Civil Aviation, pp 126–87
7. Vdovichenko VN (2001) Relationship between data transmission channel load in ATC systems with air traffic density. Abstracts of the RTS "Koncepciya sozdaniya integrirovannogo oborudovaniya, navigacii, posadki, svyazi i nablyudeniya", pp 23–24
8. Kuzmin BI, Meshalov RO (2008) Modern means of aviation telecommunications. Elektrosvyaz 5
9. Logvin AI, Vdovichenko VN (1999) Functional effectiveness of communication channels in ATC system with a random nature of radio traffic density. Abstracts of the IRTC "Sovremennyye problemy GA", pp 195–196
10. Logvin AI, Vdovichenko VN (2001) Functional effectiveness evaluation of communication channels in ATC system with a random radio traffic density. Abstracts of the IRTC "GA na rubezhe vekov", pp 170–171
11. Pestryakov VB (1983) Modern problems of radio engineering and telecommunications, problems of probabilistic and systemic approaches in research and equipment quality evaluation
12. Pestryakov VB, Kuzenkov VD (1985) Estimation of the reliability characteristics boundary values of avionic switching system functional units according to the field performance data. Radio i Svyaz, 376 pp
13. (1999) Manual of air traffic services data link applications. ICAO, second revision
14. Falkov E, Shavrin S (2017) Cyber security of aircraft information and communication systems. Inf Anal J "Radioelektronnye tekhnologii" 5

Chapter 5
Increase of Noise Immunity of Navigation and Communication Systems of Civil Aircraft During Flights in Mountainous and Urban Areas

5.1 Method of Increasing Noise Immunity During Civil Aviation Flights in Urban Areas

Since the data line from the aircraft operates within the line of sight, the errors associated with the signal propagation medium generally occur at the unfavorable terrain. For example, for the case when the transmitting antenna fixture height is 200 m, the average field strength in the urban area is 10–12 dB lower than in open space, and halving the height will increase the attenuation by 6 dB. The additional attenuation depends on the polarization of the radiation, the properties of the building materials and the location of the buildings.

Large structures, significantly larger than the wavelength of VHF radiation, create extensive shadow zones, and the scattered and reflected waves contribute to the propagation of radio waves in the city with a significantly multi-path character and form a complex interference field structure with deep and sharp spatial fading. This creates significant difficulties both in predicting the operating conditions of radio facilities and in providing reliable radio communications, especially in broadband digital communication systems and when communicating with mobile objects [1].

The polarization of the signal is changing significantly in urban conditions. Multi-path propagation and interference of radio waves cause strong variability of signal polarization.

Modern cities, in terms of radio wave propagation, represent such a complex environment that its mathematical description is inconceivable without the simplifications determined by the goals of a particular task. For VHF, large city buildings are almost opaque, and their size is many times greater than the wavelength. This leads to the formation of extensive shadow zones in the city and largely determines the properties of the emerging field.

As a model of urban development, we will take a lot of large opaque objects randomly located on a flat surface of the ground. Let us build a Cartesian rectangular coordinate system (x, y, z) by aligning the coordinate plane $z = 0$ with the ground

© The Author(s), under exclusive license to Springer Nature Singapore Pte Ltd. 2021
D. A. Zatuchny et al., *Noise Resistance Enhancement in Aircraft Navigation and Connected Systems*, Springer Aerospace Technology,
https://doi.org/10.1007/978-981-16-0630-4_5

surface. In the accepted model, the urban topography will be described by a sharply crossed surface with steep irregularities, and the surface of a continuous or being the limit of the sequence of continuous surfaces, which can be described by the equation

$$z = Z(x, y),$$

clearly defining the height of the relief z relative to the ground surface.

Let us build a characteristic function

$$\chi(r) = \begin{cases} 0, z > Z(x, y), \\ 1, z \leq Z(x, y). \end{cases}$$

At $z > 0$, we will introduce here as follows:

$$\chi(r) = \sum_{n=1}^{N} \chi_n(x, y) \chi_0^{h_n}(z),$$

where N is the number of buildings in the area; $\chi_n(x, y)$ is the function equal to one if the projection of (x, y, z) point on $z = 0$ plane falls inside or on the boundary of the n-th building, and zero otherwise; h_n is the height of the n-th building; $\chi_a^b(z)$ is the characteristic function equal to one on (a, b) interval and zero outside it.

Let us calculate the average value $\chi(r)$ from the ensemble of possible implementations. Let us consider that heights $\{h_n\}, n = 1, 2, \ldots, N$ have identical distributions for all n and do not depend on the form of projections of buildings on $z = 0$ plane, and also on their total number N. The result of averaging can be recorded as follows [2]:

$$< \chi(r) >=< \sum_{n=1}^{N} \chi_n(x, y) >< \chi_0^{h_n}(z) >,$$

because under the assumption $< \chi_0^{h_n}(z) >$ is not a function of n. Let us denote the density of the distribution of building heights through $w(h_n)$, then

$$< \chi_0^{h_n} >= \int_z^{\infty} w(h_n) \partial h_n = P_h(z)$$

and determines the likelihood of $z < h_n$. $\chi(x, y, 0) == \sum_{n=1}^{\infty} \chi_n(x, y)$ function takes unit values in $z = 0$ plane on the set, which is the union of all buildings' projections, and is equal to zero outside this set. Its average value,

$$\sum_{n=1}^{N} \chi_n(x, y) = M$$

Determines, the probability that the projection of the point on $z = 0$ plane belongs to the specified set, i.e., falls inside a building. Thus,

$$< \chi(r) >= M P_h(z).$$

If differential corrections are transmitted via the ground-to-air transmission channel, single-scattered waves make the main contribution to the formation of the received signal. Their number is a random value distributed by Poisson's law, the only parameter of which is the average number of observed reflection points $\overline{N}$, i.e. the probability of upcoming N of reflected waves is determined by the formula

$$P(N) = \frac{(\overline{N})^N}{N!} e^{-\overline{N}}. \tag{5.1.}$$

The calculated value $\overline{N}$ is approximated by the sum $\overline{N} = \overline{N_1} + \overline{N_2}$ in which the summands $\overline{N_1}$ and $\overline{N_2}$ are defined in the following formulas:

$$\overline{N_1} = \pi v d^2 (1 - \xi) \exp\left(-\frac{1}{\xi}(1 - \xi)\right) tg \frac{v_1}{2}. \tag{5.2}$$

In this formula, v is the average number of buildings per unit area,

d the distance between the aircraft and the data station;

$$\xi = (\gamma_0 d)^{-1};$$

$$\gamma_0 = 2vL/\pi,$$

where L is the average length of buildings, v is the average number of buildings per 1 km^2.

v_1 value is determined by the equation:

$$\cos v_1 = \frac{(1 + \xi)/\xi}{1 + (1 + \xi)/\xi} = \frac{1 + \xi}{1 + 2\xi}. \tag{5.3}$$

$$\overline{N_2} = \frac{\pi v}{\gamma_0^2} \frac{\xi}{1 - \xi} tg \frac{v_2}{2}. \tag{5.4}$$

v_2 value is determined by the equation:

$$tgv_2 = \frac{\pi^2}{2}(\gamma_0 d)^{-3/2} = \frac{\pi^2}{2}\xi^{3/2}. \tag{5.5}$$

Let us introduce the following value: $P_1 = 1 - e^{-\overline{N}}$. P_1 value determines the probability of the arrival of single reflected waves at the observation point [3]. P_1 value should be considered as the lower limit when estimating the probability of arrival of single reflected waves. The dependencies of P_1 probability of the distance between the aircraft and the data reception point are given with the values of the aircraft height equal to 300 m (curve 1), 400 m (curve 2), 500 m (curve 3), and 600 m (curve 4). The height of the urban development layer is assumed to be 20 m; $\gamma_0 = 6\,km^{-1}$.

As it can be seen from Fig. 5.1, the probability of reflected waves arrival decreases with the increase in the distance between the aircraft and the data transfer point. This function is nonlinear. It should be noted that at a fixed distance between the aircraft and the data point, the probability of reflected waves increases with the height of the aircraft flight.

As mentioned above, the main contribution to the signal formation is made by the scattered waves once. Let us draw a conclusion [4]:

$P_1 \to 0$, if $\overline{N} \to 0$.

It follows from (5.2) that this event occurs if

$$v \to 0,$$
$$d \to 0,$$
$$\xi \to 0,$$
$$v_1 \to 0.$$

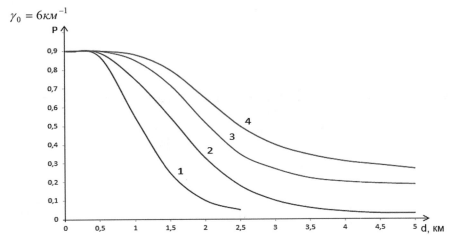

Fig. 5.1 Relationship between the probability of receiving reflected waves and the distance between the aircraft and the data reception point

Since $\exp\left(-\frac{1}{\xi}(1-\xi)\right)$ value decreases faster than the v and d values, and the conditions $\xi \rightarrow 0$ and $v_1 \rightarrow 0$ are equivalent, let us propose the condition of obtaining the minimum $\overline{N}_1$ value:

$$
\begin{aligned}
&\xi \rightarrow 0, \\
&d \le d_1, \\
&v \le V_1,
\end{aligned}
\tag{5.6}
$$

where d_1, V_1 are some admissible values related to the conditions of a particular city. It follows from (5.4) that

$$
\begin{aligned}
&v \rightarrow 0, \\
&\frac{\xi}{1-\xi} \rightarrow 0, \\
&v_2 \rightarrow 0.
\end{aligned}
$$

Since the $\xi \rightarrow 0$ condition follows from the $\frac{\xi}{1-\xi} \rightarrow 0$ condition, and $v_2 \rightarrow 0$ and $\xi \rightarrow 0$ conditions are equivalent, let us propose the condition of obtaining the minimum $\overline{N}_2$ value [4, 5]:

$$
\begin{aligned}
&\xi \rightarrow 0, \\
&v \le V_2,
\end{aligned}
\tag{5.7}
$$

where V_2 is some admissible value related to the conditions of a particular city.

The criterion for choosing a data transmission location to mitigate the effects of wave reflection follows from conditions (5.6) and (5.7):

$$
\begin{aligned}
&\xi \rightarrow 0, \\
&d \le d_1, \\
&v \le V,
\end{aligned}
$$

where $V = \min(V_1, V_2)$.

Measurements have shown that the signal strength decreases back proportionally to the third degree of the distance at distances up to 15–20 km. Subsequent increase in range leads to a faster decrease in signal strength. In radial streets, the signal level is 10–15 dB higher than in transverse streets in relation to the route. However, this ratio depends on the distance: the difference decreases as the distance increases. It should be noted that as the distance increases, the role of the wave propagating over urban development increases. Its contribution becomes decisive at distances exceeding 1.5–2 km and depends on the angles of aircraft shelter and the data reception point created by close-knit buildings [6].

Typically, the reflective surfaces of buildings are significantly heterogeneous due to the presence of windows, balconies, and other elements that also have different electrical properties. Approximately reflecting surfaces of buildings are described by a complex reflection coefficient with a phase evenly distributed in $[0, 2\pi]$ interval and correlation scales in horizontal l_Γ and vertical l_B directions. Thus, building surfaces can be considered as amplitude-phase screens, which are widely used to describe the scattering process on statistically heterogeneous building surfaces. The actual surface itself is not continuous, but is composed of many amplitude-phase screens of finite dimensions, randomly and independently located on the surface of the ground, i.e. the scattering surface is a set of discrete statistically independent amplitude-phase screens.

Let the relief of urban development be described by some random surface S. The field above the surface according to Green's theorem can be represented as an integral equation:

$$U(r_2) = U_i(r_2) + \int_S \left\{ U(r_s) \frac{\partial G(r_2, r_s)}{\partial n_s} - G(r_2, r_s) \frac{\partial U(r_s)}{\partial n_s} \right\} \partial S, \qquad (5.8)$$

where $U_i(r)$ is the falling field; G—Green's function; n_S—normal to the surface S at r_S point.

Surface S is formed by a horizontal flat surface of the ground S_1 and a random surface S_2, which is the combined surface of many urban objects, orthogonal to S_1 and located on it in a random way. Let us build the Green's function $G(r_2, r_1)$ as follows:

$$G(r_2, r_1) = \frac{1}{4\pi} \left\{ \frac{e^{ik|r_2 - r_1|}}{|r_2 - r_1|} - \frac{e^{ik|r_2 - r_1'|}}{|r_2 - r_1'|} \right\}, k = \frac{2\pi}{\lambda}, \qquad (5.9)$$

where r_1' is the point mirroring the r_1 from the ground surface; S_1, λ is the wavelength of radiation.

Assuming that the wave emitted by the source located at r_1 point is described by taking into account the influence of the earth by the formula (5.9), we transform the ratio (5.8) to the form:

$$U(r_2) = G(r_2, r_1) + \int_S \left\{ U(r_s) \frac{\partial G(r_2, r_s)}{\partial n_s} - G(r_2, r_s) \frac{\partial U(r_s)}{\partial n_s} \right\} \partial S. \qquad (5.10)$$

In (5.10), the Green's function is defined by the formula (5.9) and thus excludes integration by S_1. The boundary values $U(r_S)$ and $\frac{\partial U(r_s)}{\partial n_s}$ include the contribution of $U_i(r_S)$ element falling on ∂S element and reflected (scattered) $U_r(r_S)$ emission. Since the object of urban development is opaque, the radiation $U_i(r_s)$ that comes only from the part of the half-space where the aircraft is located may fall on its side facing the aircraft. Integral to (5.10) has such a value, that in this case

$$\int_{S_2} \left\{ U_i(r_s)\frac{\partial G(r_2, r_s)}{\partial n_s} - G(r_2, r_s)\frac{\partial U_i(r_s)}{\partial n_s} \right\}\partial S = 0.$$

$U_r(r_s)$ refers to radiation scattered on the side of the screen facing the aircraft. Thus,

$$U(r_2) = G(r_2, r_1) + \int_{S_2} \left\{ U_r(r_s)\frac{\partial G(r_2, r_s)}{\partial n_s} - G(r_2, r_s)\frac{\partial U_r(r_s)}{\partial n_s} \right\}\partial S,$$

or this ratio can be represented in the form of

$$U(r_2) = G(r_2, r_1) + 2\int_{S_2} U_r(r_s)(n_s\nabla_s)G(r_2, r_s)\partial S,$$

where $\nabla_s = \left(\frac{\partial}{\partial x_s}, \frac{\partial}{\partial y_s}, \frac{\partial}{\partial z_s} \right)$.

In Fig. 5.2, the geometry of the problem of calculation of once scattered waves is shown.

Fig. 5.2 Geometry of the problem of calculation of once scattered waves

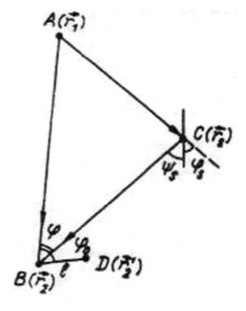

$U_r(r_s)$ value is defined as a product of the coefficient of reflection $R(\varphi_s, r_s)$ on the boundary value of the incident radiation $U_i(r_s)$ and on the shading function $Z(r_2, r_s, r_1)$, which is equal to one if r_s point is visible from the points r_2 and r_1 simultaneously, and zero in all other cases. In approximation of single scattering, we will write down the last formula in the following form [2, 7]:

$$U(r_2) = Z(r_2, r_1)G(r_2, r_1) + 2\int_{S_2} Z(r_2, r_s, r_1)R(\varphi_s, r_s)G(r_s, r_1)(n_s \nabla_s)G(r_2, r_s)\partial S,$$

at that $\sin \varphi_s = \left(n_s \frac{r_s - r_1}{|r_s - r_1|}\right)$

Let us write the following:

$$(n_s \nabla_s)G(r_2, r_s) \approx ik\left(n_s \frac{r_2 - r_s}{|r_2 - r_s|}\right)G(r_2, r_s) = ik \sin \psi_s G(r_2, r_s).$$

Taking into account the last ratio, let us rewrite the expression for $U(r_2)$ field in the form

$$U(r_2) = Z(r_2, r_1)G(r_2, r_1)$$
$$+ 2ik \int_{S_2} Z(r_2, r_s, r_1)R(\varphi_s, r_s) \sin \psi_s G(r_2, r_s)G(r_s, r_1)\partial S.$$

Suppose that the reflective properties of buildings are random and independent, but statistically the same; the reflection coefficient value is complex, and the phase with equal probability can take any value on $[0, 2\pi]$ interval, so that the average value is $< R(\varphi_s, r_s) >= 0$. This leads to the fact that after averaging the reflection properties and position of all buildings, the following expression may be written for $U(r_2)$:

$$< U(r_2) >=< Z(r_2, r_1) > G(r_2, r_1) = P_{21}G(r_2, r_1),$$

where P_{21} is the probability of receiving a signal directly from the aircraft.

Therefore, the average value of the field at the receiving point during the propagation of radio waves in city conditions will be defined by the probability of receiving the signal directly from the aircraft.

The average intensity of the field above urban development is equal to the sum of the intensity $< I >$ of the scattered waves and the intensity of the coherent wave $< I_k >$ coming directly from the aircraft, the calculation of which should take into account the probability of direct visibility between the aircraft and the data reception point.

Within the framework of a statistically homogeneous urban area model with an average building height h, the probability of a line of sight between the aircraft located at the coordinates $r_1(x_1, y_1, z_1)$ and the data reception point at the point $r_2(x_2, y_2, z_2)$ at $0 < z_1, z_2 < h$ is defined by the expression

$$P(r_1, r_2) = \exp\left(-\gamma_0\sqrt{(x_2 - x_1)^2 + (y_2 - y_1)^2}\right),$$

where

$$\gamma_0 = 2\nu L/\pi,$$

where L is an average length of buildings, ν is an average number of buildings per unit area

At $0 < z_2 < h$ and $z_1 > h$,

$$P(r_1, r_2) = \exp\left(-\gamma_0\sqrt{(x_2 - x_1)^2 + (y_2 - y_1)^2}\frac{h - z_2}{z_1 - z_2}\right).$$

Table 5.1 presents the dependence of the line-of-sight probability on the average number of buildings per unit area. When calculating, the average length of buildings was assumed to be 20 m, the average height of buildings was assumed to be 20 m, the height of the data reception point was 10 m, $x_2 - x_1 = 10$, $y_2 - y_1 = 10$..

As it can be seen from the table above, the probability of the line of sight decreases as the number of buildings per unit area increases. However, this trend is increasing with the increase in the number ν. If the value decreased by 2.46 times during the transition from $\nu = 0.5$ to the $\nu = 1$, the value decreased by 222.27 times during the transition from $\nu = 3$ to $\nu = 6$.

Based on the analysis of the results of Table 5.1. The following conclusion can be made: when placing the data reception point, it is necessary to take into account the number of buildings in the selected area: it is desirable to choose the location of this point in the area with the smallest development [8].

Table 5.1 Dependence of the line-of-sight probability on the average number of buildings per unit area

ν	$P(r_1, r_2)$
0.1	0.835
0.5	0.406
1	0.165
2	0.027
3	$4.49 \cdot 10^{-3}$
4	$7.42 \cdot 10^{-4}$
5	$1.23 \cdot 10^{-4}$
6	$2.02 \cdot 10^{-5}$

Table 5.2 The probability of the line of sight as a function of the aircraft flight altitude

z_1	$P(r_1, r_2)$
$\nu = 1$	
500	0.025
700	0.074
1000	0.165
1200	0.221
1500	0.298
2000	0.402
$\nu = 0.1$	
500	0.694
1000	0.835
2000	0.917

Table 5.2 presents the dependence of the line-of-sight probability on the aircraft flight altitude. The average length of buildings was assumed to be 20 m, the average height of buildings was assumed to be 20 m, the height of the data reception point was 10 m, $x_2 - x_1 = 10$, $y_2 - y_1 = 10$, $\nu = 1$ and $\nu = 0.1$.

As it can be seen from Table 5.2, the probability of the line of sight increases with the increase in the altitude of the aircraft flight. This trend slows down as z_1 value increases. If the value increased by 6.6 times during the transition from $z_1 = 500$ to $z_1 = 1000$, the value increased by 2.44 times during the transition from $z_1 = 1000$ to $z_1 = 2000$.

The probability of the line of sight as a function of distance between the ground plane projection of the aircraft and data acquisition facility location points is shown in Table 5.3. For the calculation goals, the average length of buildings was assumed to be 20 m, the average height of buildings was assumed to be 20 m, the height of the data reception point was 10 m, the height of the aircraft 1,000 m.

As can be seen from Table 5.3, the probability of the line of sight decreases with increasing distance between projections of points on the ground plane. This trend is increasing as $\sqrt{(x_2 - x_1)^2 + (y_2 - y_1)^2}$ value increases. If this value decreases by

Table 5.3 The probability of the line of sight as a function of the distance between the ground plane projection of the aircraft and data acquisition facility location points

$\sqrt{(x_2 - x_1)^2 + (y_2 - y_1)^2}$	$P(r_1, r_2)$
0.5	0.938
1	0.880
2	0.773
4	0.597
8	0.357
16	0.128

1.066 times during the transition from 0.5 to 1, the value decreases by 2.789 times during the transition from 8 to 16.

Based on the analysis of Tables 5.2 and 5.3, we will propose a criterion for selecting the point of the aircraft route for transmitting information to the data reception point located in the city [6, 9]:

$$P(r_1, r_2) \to \max,$$
$$h \geq \alpha,$$
$$\sqrt{(x_2 - x_1)^2 + (y_2 - y_1)^2} \leq \beta,$$

where α, β are some values.

In Figs. 5.3, 5.4 and 5.5 the graphical type of dependencies given in Tables 5.1, 5.2 and 5.3 is shown.

Let us draw a conclusion: the main contribution to the accuracy and reliability of the navigation support of the aircraft over the layer of urban development is the distance between the aircraft and the data reception point. Thus, it is necessary for the

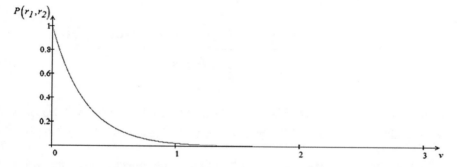

Fig. 5.3 The probability of the line of sight as a function of the average number of buildings

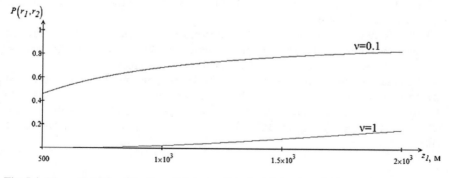

Fig. 5.4 The probability of the line of sight as a function of the aircraft flight altitude

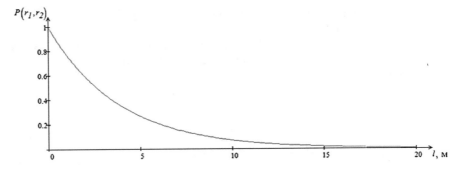

Fig. 5.5 The probability of the line of sight as a function of distance between the ground plane projection of the aircraft and data acquisition facility location points

aircraft crew to reduce the distance to the point of transfer of differential corrections as much as possible (to the extent that urban conditions allow), while at the same time making sure that the aircraft's flight altitude does not fall below 500 m.

The dependence of the average field intensity on the aircraft flight height is described by the function

$$f(k) = \frac{\left| 1 + k\sqrt{2\pi}\exp\!\left(i\left(\frac{\pi}{4} + 2k^2\right)\right)\left[1 + erf\left(k\sqrt{2}\exp\left(i\frac{\pi}{4}\right)\right)\right]\right|}{k + \left(\frac{k^2 + \sqrt{k^4+1}}{2}\right)^{\frac{1}{2}}},$$

where $k = \frac{\sqrt{\pi}}{2}\frac{z_1 - h}{(\lambda \tilde{r}/2)^{\frac{1}{2}}}$, where $\tilde{r}$ is $|r_2 - r|$ projection on the plane $z = 0$, $erf(x)$ is the integral of the probability. The graph of $f(k)$ function is shown in Fig. 5.6 (curve 1).

This expression is cumbersome and is not quite convenient for analysis. Let us use $f(k)$ instead of its approximation $\widetilde{f}(k) = \sqrt{2 + (4\pi k)^2}$, which coincides with $f(k)$, and $k = 0$ and $k \to \infty$ in the extreme cases. The graph of $\widetilde{f}(k)$ function is shown in Fig. 5.3 (curve 2).

When the height of the aircraft's flight decreases, the diffraction of the scattered field over the building layer starts to play a significant role and leads to the fact that $< I >$ does not turn to zero at $z_1 = h$ (dashed line in Fig. 5.3).

At a small difference of heights $(z_1 - h)$, the influence of the curvature of the ground surface on the character of shading generated by buildings is noticeable already at a distance of several kilometers from the aircraft. This fact can be taken into account by introducing the following value instead of $(z_1 - h)$ difference [3, 10]:

$$H = (z_1 - h) - \frac{d^2}{2R}, H > 0,$$

Fig. 5.6 The dependence of mean wave intensity on the height of the observation site

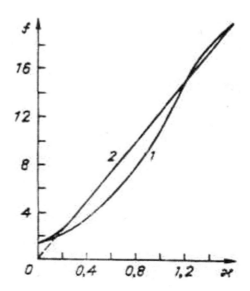

where $R = 6370$ km is the Earth's radius, d is the distance between the aircraft and the data reception point.

However, the introduction of H value is valid only for distances shorter than the radio-horizon range d_R. It is necessary to take into account the diffraction of waves on the spherical surface of the Earth and the scattering of waves on inhomogeneities of the troposphere with $d > 0.8d_R$. Figure 5.7 qualitatively illustrates the contributions of these mechanisms to the formation of the field beyond the horizon. A solid line in the figure shows the relative attenuation of the diffraction wave field, and a dashed line shows the relative attenuation of the scattering wave field.

At distances of less than 100 km, the main contribution is made by the effects of diffraction, and wave scattering in the atmosphere can be disregarded. Since the distance from the aircraft to the receiving point does not actually exceed 100 km, it is necessary to propose a method of calculating the diffraction field.

Introduces the concept of "scale of distances" and "scale of heights" as defined by the ratios

$$L_1 = \left(\lambda R^2/\pi\right)^{\frac{1}{3}} \text{ and } H_1 = \tfrac{1}{2}\left(\lambda^2 R/\pi^2\right)^{\frac{1}{3}}.$$

The following values are entered:

$$x = \frac{d}{L_1},$$

where

$$y_1 = \frac{z_1}{H_1}, \, y_2 = \frac{z_2}{H_1}.$$

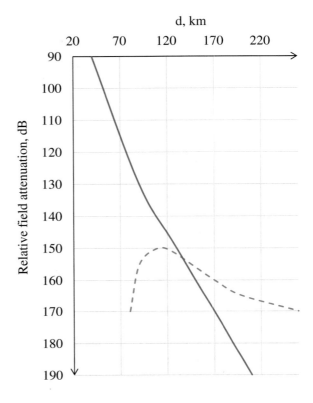

Fig. 5.7 Contribution of wave diffraction on the spherical surface of the Earth and wave scattering on troposphere inhomogeneities to the formation of the field beyond the horizon

The attenuation multiplier looks as follows:

$$F = 2\sqrt{\pi\,x}\left|\sum_{s=1}^{\infty}\frac{e^{ixt_s}}{t_s+q^2}\frac{h_2(t_s+q_1)}{h_2(t_s)}\frac{h_2(t_s+y_2)}{h_2(t_s)}\right|,$$

where $q = i\,(\pi\,R/\lambda)^{1/3}/\sqrt{\varepsilon_k}$; ε_k is a complex dielectric permeability of the earth; $h_2(t)$ is Airy function, t_s are the roots of the equation

$$h'_2(t) - qh_2(t) = 0.$$

In the shadow zone, the number of lines quickly converges and can be limited only to the first term, which has a clearly defined structure of the species

$$F = U(x)V(y_1)V(y_2),$$

the first multiplier of which depends only on the distance:

$$U(x) == 2\sqrt{\pi}\,x\left|\frac{e^{ixt_1}}{t_1 + q^2}\right|,$$

and the second and third are from aircraft heights and data reception points, respectively, and are called height multipliers:

$$V(y) = \left|\frac{h_2(t_1 + y)}{h_2(t)}\right|,$$

at that, if the antenna is at a ground level, then $y = 0$ and the height factor $V(0)$ is equal to one.

The given formulas allow to calculate the field weakening due to diffraction on the spherical surface of the Earth both in the area of its geometrical shade and in the area of a penumbra. However, where high accuracy is not required, simpler and more convenient calculation methods can be used to calculate the attenuation factor F. One of them is a Bullington's nomogram in Fig. 5.8.

For the communication line with *l*/length shown in Fig. 5.9. First determine the "illumination" zones with respect to the height of the aircraft and the data reception point according to the formula:

$$l_i = \sqrt{mz_i}, \quad i = 1, 2,$$

where m is the ratio of the effective radius of the Earth to the true radius R, which equals to approximately $\frac{4}{3}$. l_1 always indicates the smallest of the distances, i.e. $l_1 < l_2$. The shadow zone l_3 is found from the ratio

$$l_3 = l - l_1 - l_2.$$

Then according to the nomogram depicted in Fig. 5.8. There is a field attenuation coefficient for each of the zones l_1, l_2, l_3.

The mentioned field calculation methods can be used to calculate the average intensity of a VHF field above the urban development layer, which, considering the phenomenon of wave diffraction on a spherical surface of the Earth, can be represented as follows [8, 9]:

$$< I(r_2) >= \frac{\Gamma}{8\pi} \frac{\lambda l_B}{\lambda^2 + [2\pi l_B \gamma_0(z_1 - h)]^2} \frac{h - z_2}{d^3} F. \tag{5.11}$$

From (5.11), there follows the criterion of data transmission from the aircraft:

$$d \rightarrow \min,$$
$$F \leq \gamma$$

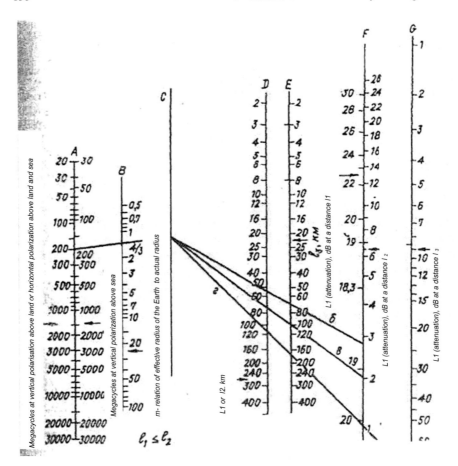

Fig. 5.8 Nomogram of signal attenuation during radio waves propagation over the smooth spherical surface of the Earth

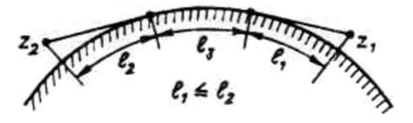

Fig. 5.9 To shadow zones determination

where γ is a certain constant.

To show the influence of the urban development profile on the field decay nature, consider the following profile:

$$P_h(z) = \chi(h_1 - z) + \chi(z - h_1)\left(\frac{h_2 - z}{h_2 - h_1}\right)^n, z \in (0, h_2), n > 0.$$

The formula $\chi(x)$ designates the Heaviside step function; h_1 and h_2 are minimum and maximum heights of buildings, respectively. The form $P_h(z)$ depending on n is shown in Fig. 5.10.

When $n \gg 1$, $P_h(z)$ corresponds to the case when high buildings are relatively rarely dominating above the urban development area with the height of approximately h_1, and at $n \ll 1$, it corresponds to the case of high-rise buildings within the urban development area with the height mainly close to h_2. In extreme cases ($n \to 0$ and $n \to \infty$) $P_h(z)$ corresponds to the model of a single-level building, and in the first case, all buildings have a height of h_2, and in the second case, height of h_1.

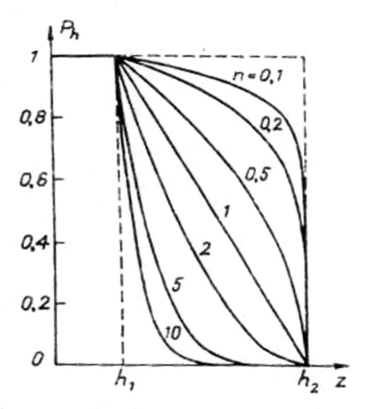

Fig. 5.10 $P_h(z)$ view as a function of n

Thus, the selected representation describes a wide class of urban high-rise models: from s single level to a model with an equal height of buildings in the range from h_1 up to h_2 at $n = 1$.

5.2   Increase of Noise Immunity of Navigation and Communication Systems of Civil Aircraft During Flights in Mountainous and Urban Areas

Inability of VHF waves to bend around the earth's surface requires geometric visibility between the transmitting and receiving antennas (Figs. 5.11 and 5.12) to provide communication.

Since the waves are reflected from the earth's surface at the reception site, as we can see from Fig. 5.13, ray interference may happen; as a result, interference fading and distortion of transmitted messages occur. The influence of ground irregularities and soil inhomogeneity on the lower air layers, the difference and, accordingly, the

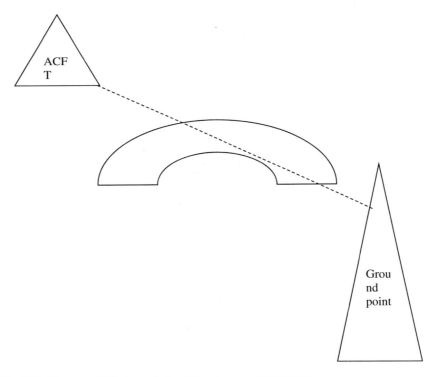

Fig. 5.11   The case of data transmission failure in mountainous areas

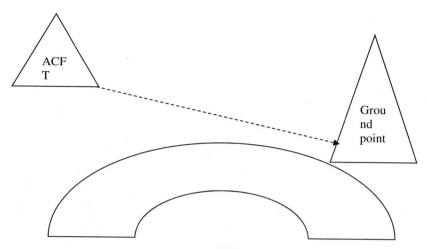

Fig. 5.12 The case of data transmission in mountainous areas

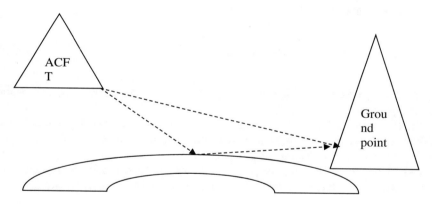

Fig. 5.13 Ray interference

unequal influence of vegetation cover in certain areas of the territory of waves propagation, presence of rivers and water reservoirs, as well as settlements and engineering structures, etc., lead to the formation of atmosphere zones with various temperature and humidity, local air flows, etc. In these zones formed at altitudes up to several kilometers, wave diffusion occurs, as it is schematically shown in Fig. 5.14. In this case, the part of the wave energy reaches points spaced from the transmitting antenna by a distance several times greater than the range of geometric visibility.

With a significant transmitter power, sharply directed antennas and a receiver with high sensitivity, the scattering of waves in tropospheric irregularities at altitudes of 2–3 km allows to receive radio communication at distances of hundreds of kilometers, which is 5–10 times greater than the distance of geometric visibility.

The line-of-sight distance can be calculated with the equation [11]:

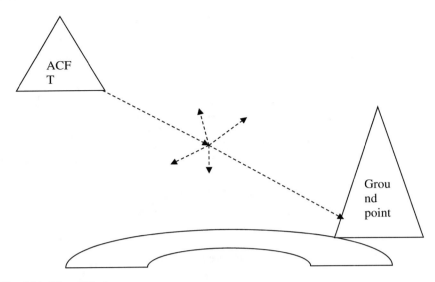

Fig. 5.14 Wave diffusion

$$R_0 = 4.12\left(\sqrt{h_1} + \sqrt{h_2}\right), \tag{5.12}$$

where R_0 is a line-of-sight distance, h_1 is the transmitting antenna height, h_2 is the receiving antenna height.

Table 5.4 shows data on the change of the line-of-sight distance depending on the altitude of the aircraft. During calculation, it was assumed that $h_2 = 2000$ m. Data on R_0 are given in km, data on h_1 are given in.

Field intensity at the reception point within free space, i.e. in the case when there are no obstacles in the radio waves path, can be calculated with the equation:

$$E_0 = 222 \cdot 10^3 \sqrt{\frac{P_\ni}{R}}, \tag{5.13}$$

Table 5.4 Data on the change of the line-of-sight distance depending on the altitude of the aircraft

h_1	R_0
2.5	390.25
3	409.90
3.5	427.99
4	444.84
4.5	460.62
5	475.57

Table 5.5 Data on the dependence of the field intensity on the distance within open space for various values of the effective radiated power

R	E_0
$P_9 = 0{,}1$	
1	70,202
5	31,395
10	22,200
20	15,698
30	4053
50	3140
$P_9 = 1$	
1	222,000
5	99,281
10	70,202
20	49,640
30	40,531
50	31,395
$P_9 = 10$	
1	702,025
5	313,955
10	222,000
20	156,977
30	128,171
50	99,281

where E_0 is the field intensity, μV/m; P_9 is the effective radiated power, kW, R is the distance, km.

Table 5.5 shows data on the dependence of the field intensity on the distance within open space for various values of the effective radiated power.

One of the factors affecting data transmission in mountainous regions is multi-path propagation due to the reflections from the ground surface. Errors due to multi-path propagation are difficult to assess due to their strong dependence on the type and relative position of the antennas, the type and nature of the reflecting surface.

Direct wave and the wave reflected from the ground surface get to the acquisition facility. If these wave phases coincide, the field intensity doubles; when they are antiphasic, then the intensity decreases down to zero.

The phase ratio between the direct and reflected waves is determined by their path difference and the fact that when reflected from a conducting surface the wave phase is inverted. Field alteration due to the reflection influence can be calculated with the equation [11]:

Table 5.6 Data on the field	R	Y (dB)
alteration depending on the	5000	−41
distance between the aircraft	3000	−39
and the data acquisition	1000	−34
facility	500	−31

$$Y = 10 \lg\left[4 \sin^2(2\pi\, h_1 h_2)/(R\lambda)\right], \qquad (5.14)$$

where Y is the field intensity alteration due to reflections, dB; R is the distance, m; λ is the wavelength, m; h_1, h_2 is the aircraft altitude; and the ground data acquisition facility, m.

From (5.14), it follows that the distance between the aircraft and the data acquisition facility as well as the data transmission wavelength has the primary influence on the field alteration due to the influence of reflection.

Table 5.6 shows data on the field alteration depending on the distance between the aircraft and the data acquisition facility. The following data were taken into account in the calculations: the altitude of the aircraft flight $h_1 = 3000$ m, the height of the data reception point $h_2 = 2000$ m.

Let us introduce the following symbols [12]:

$$4 \sin^2(2\pi\, h_1 h_2) = y_1,$$

$$R\lambda = y_2.$$

Let us note that $Y \to \max$ if $y_1 \to \max$, $y_2 \to \min$.

We should also note that $y_1 \to \max$, if $2\pi\, h_1 h_2 = \frac{\pi}{2} + 2\pi\, k$ or $h_1 h_2 = \frac{1}{4} + \frac{k}{2}$, where k is an integer number.

Let us propose a criterion for the aircraft flight altitude selection and the distance to the ground acquisition facility for minimal field change due to reflection effect:

$$R\lambda \to \min,$$

$$h_1 h_2 = \frac{1}{4} + \frac{k}{2},$$

where k is an integral number.

The criterion includes a recommendation to the pilot of the aircraft: reduce the distance to the data acquisition facility during data transmission as far as the terrain conditions and the flight situation allow, while adhering to the flight altitude that satisfies the above restriction.

The reflected radio waves enter the receiving point with a Δt delay determined by the difference in travel L [13]:

$$\Delta t = \frac{L}{c}, \tag{5.15}$$

where c is radio wave velocity equal to the speed of light; L—travel difference.

The specific character of multi-path signal propagation consists in dependence of the reflected signal amplitude from its delay. This function is as follows:

$$U = \begin{cases} 0 & \tau^* - \tau_3 > T_\kappa, \\ \left(1 - \dfrac{\tau^* - \tau_3}{T_\kappa}\right) U_{отр}, & \tau^* - \tau_3 \leq T_\kappa, \end{cases} \tag{5.16}$$

where τ^* is the estimate of the delay time, T_κ is the data sequence autocorrelation function width, τ_3 is the delay time of the reflected signal, U_{OTP} is the amplitude of the reflected signal.

According to formula (5.16), the reflected signals, with a delay greater than T_κ relative to the direct signal, shall not affect data transmission via data transmission line.

As it follows from (5.15), the difference in the travel path affects the reflected radio wave delay. In this case, taking into account (5.16), the following ratio shall be fulfilled:

$$L > cT_k \tag{5.17}$$

An example of multi-path wave propagation is shown in Fig. 5.15.

Let us draw a conclusion: in case of multi-path radio waves propagation, it is necessary to ensure that data acquisition facility receives only those waves, the travel path difference of which fulfills the condition (5.17). To fulfill this condition, it is necessary, in turn, to fulfill the requirements related to the aircraft flight altitude, signal propagation angle, and location of the data acquisition facility [14].

To calculate the errors associated with multi-path nature, it is necessary to build a mathematical model of the mountainous region. Consider the region of the Caucasus Mountains. This is a high-mountain region that contains a number of single ridges; the area of the region is 43 thous. km^2. Mathematical modeling is performed by an overlay method. Mountain structures are described by the following points: the vertex coordinates and the base coordinates of the facets. It is assumed that all the facets start from the summit and go down to the base. The current height of the mountain structure within this model is described by the formula [12]:

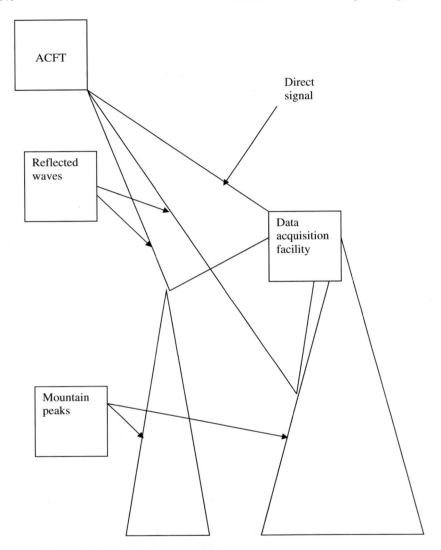

Fig. 5.15 Example of multi-path wave propagation

$$
h(x, y)_{|hm,\Delta x_1,\Delta x_2,\Delta y_1,\Delta y_2} =
$$

$$
\begin{cases}
\dfrac{h_m}{\sqrt{\left[1+\left(\frac{y}{\Delta y_{1(2)}}\right)^2\right]\left[1+\left(\frac{x}{\Delta x_1}\right)^2\right]}}, & x \geq 0, y \geq 0(y \leq 0) \\[4ex]
\dfrac{h_m}{\sqrt{\left[1+\left(\frac{y}{\Delta y_{1(2)}}\right)^2\right]\left[1+\left(\frac{x}{\Delta x_2}\right)^2\right]}}, & x < 0, y \geq 0(y < 0)
\end{cases}
$$

$$(5.18)$$

where h_m is the maximum height of the mountain structure, Δx_1, Δx_2 is the width of the mountain structure along the X-axis at a level $h_m\sqrt{2}$ in the positive (Δx_1) and negative (Δx_2) directions, Δy_1, Δy_2 is the width of the mountain structure along the Y-axis. The height values and parameters of the mountain structure bases are determined by the physical map. In addition to the coordinates of the summit and base points, it is necessary to know the angle of rotation of Cartesian axes associated with this structure in the Cartesian frame of reference associated with the considered region.

The result is eight parameters that completely describe the mountain structure. This can be represented as follows:

$$h_i(x, y) = h(u, v)|_{h_{mi}, \Delta x_{1i}, \Delta x_{2i}, \Delta y_{1i}, \Delta y_{2i}},$$

where $u = (x - x_{Bi})\cos\psi_i + (y - y_{Bi})\sin\psi_i$, $v = (x_{Bi} - x)\cos\psi_i + (y - y_{Bi})\sin\psi_i$, i is the number of mountain structure, x_{Bi} and y_{Bi} is the coordinates of the ith summit, ψ_i is the angle of rotation of Cartesian axes associated with the mountain structure around the Z-axis in the Cartesian frame of reference associated with the considered region.

The mathematical model of the mountainous region was entirely formed by combining separate mountain structures according to the following algorithm:

$$h_i(x, y) = \max\{h_i(x, y)\}, i = 1 \ldots N,$$

where N is the number of mountain structures.

The combination of mountain structures into a single mountain area according to this algorithm allows obtaining a model of the selected mountainous terrain region.

The attenuation coefficient for diffraction is calculated using the ray optics methods, the applicability of which in this case is determined by the wavelength dimensions, which are smaller than the mountainous formations dimensions.

The attenuation factor relative to free space is calculated by the formula [15]:

$$F = \sqrt{\frac{C^2(\mu) + S^2(\mu)}{2}}, \tag{5.19}$$

where $C(\mu) = \frac{1}{2} - \int_0^\mu \cos\frac{\pi x^2}{2}\partial x$, $S(\mu) = \frac{1}{2} - \int_0^\mu \sin\frac{\pi x^2}{2}\partial x$, $\mu = \frac{h\sqrt{2}}{b}$, h is the obstacle height, $b \approx \sqrt{l\,\lambda}$, l is the distance from the reflecting surface to the aircraft, λ is the wavelength.

Table 5.7 shows the dependence of the attenuation factor on the rock formation height. The distance to the reflecting surface was taken equal to 500 m for the calculations.

Table 5.8 shows the dependence of the attenuation factor on the distance from the reflecting surface to the aircraft. The height of the obstacle was taken equal to 500 m in the calculations [16].

Table 5.7 Dependence of the attenuation factor on the rock formation height

H (m)	F
500	7.73
1000	17.42
1500	25.46
2000	35.46
2500	41.62
3000	58.66

Table 5.8 Dependence of the attenuation factor on the distance from the reflecting surface to the aircraft

l (m)	F
500	7.73
1000	6.73
1500	4.55
2000	4.66
2500	3.85

Considering Tables 5.7 and 5.8, we can draw the following conclusions:

(1)   The attenuation factor with respect to free space increases in proportion to the height of the mountainous obstacle.
(2)   The attenuation factor with respect to free space decreases when the distance to the reflecting surface increases, but more slowly than the increase of this indicator value.

Consequently, we propose a criterion for data transmission from the aircraft:

$$h \rightarrow \min,$$
$$l \geq L,$$

where L is a certain limiting value.

According to the proposed criterion, considering the data given in Tables 5.7 and 5.8, the data acquisition facility shall be located behind the mountainous formations at the altitude of 500 m maximum, while the distance to this obstacle shall be 2500 m minimum.

References

1. Akinshin RN, Zatuchnyy DA, Shevchenko DV (2016) Reducing the multi-path propagation impact when transmitting data from the aircraft. Informatizaciya i svyaz 2018(3):6–12
2. Zatuchnyy DA, Logvin AI (2013) Criteria for reducing dynamic errors in ADS mode. Nauchniy Vestnik of MSTUCA 189:5–8
3. Zatuchnyy DA, Logvin AI, Nechayev EE (2012) Problems of ADS mode implementation in the Russian Federation. Printing and publication department MSTU CA
4. Zatuchnyy DA, Slad ZhV (2015) About the impact of build-in area type on radio waves propagation in the city. Nauchniy Vestnik of MSTUCA (222):37–43
5. Karyukin GE (2005) Improvement of SRNS signal processing algorithm in multi-path propagation reception condition. Interuniversity collection of scientific works "Problemy ekspluatacii i sovershenstvovaniya transportnyh sistem", vol XI. Academy of CA, St. Petersburg
6. Kozlov AI, Sergeyev VG (1998) Radio waves propagation along natural paths. MSTUCA
7. Pestryakov VB (1990) Avionic equipment engineering. Sov Radio
8. Required navigation performance (RNP) Manual, Third Revision (2008)
9. Sizyh VV, Shahtarin BI, Shevcev VA (2017) Cycle slip mechanism in stochastic analog systems of first- and second-orders of phase-lock. Mekhatronika, avtomatizaciya, upravlenie 18(1)
10. Tikhonov VI (1982) Statistical radio engineering. Radio i Svyaz 624 pp
11. Tikhonov VI, Kulman NK (1975) Non-linear filtering and quasi-coherent signal reception. Sov Radio 704 pp
12. Zatuchnyi DA (2012) Reducing errors in data transmission from the aircraft in mountainous areas for the VHF band by improving the transfer point selection. Nauchniy Vestnik of MSTUCA (176):150–153
13. Tikhonov VI, Kharisov VN (1991) Statistical analysis and synthesis of radio-technical devices and systems. Radio i Svyaz 608 pp
14. Filin AD, Shatrakov YuG, Yakovlev VT, Yushkov AV (2017) The impact of flight support subsystems reliability on probability of aircraft accidents. Nauchniy Vestnik of Russian military-industrial complex 3:68–74
15. Kharin EG (2002) Complex information processing of aircraft navigation systems. Moscow
16. Shahtarin BI, Aslanov TG (2014) Average time to cycle slip in continuous and discrete automatic phase-lock. Vestnik of the Moscow State Technical University named after N.E. Bauman, No. 1(52)

Appendix A

See Table A.1.
    See Table A.2.
    See Table A.3.
    See Table A.4.
    See Table A.5.
    See Table A.6.
    See Figs. A.1, A.2, A.3, A.4.

Table A.1 The list of possible goals of the intruder

Item no.	Name	Action	Nos. of delivered threats	Notes
1	Breach of confidentiality of information	Unauthorized copying of confidential information or any other security leakage	1, 2, 4, 5, 6, 7, 8, 9, 13, 15, 16, 17	
2	Downgraded service	Deterioration of temporal or other parameters of user requests servicing process	1, 3, 10, 11, 12, 13, 14, 16, 17	
3	Unauthorized modification of information	Unauthorized deletion or modification of confidential information	1, 2, 3, 6, 7, 8, 9, 13	
4	Malfunction of system or individual elements thereof	Transition of the system in the inoperable state or degrading its operating parameters below the allowable limit	1, 2, 3, 10, 11, 12, 14, 15, 16, 17	

Table A.2 List of information threats

Item no.	Name	Source of threat	Possible ways of implementation	Threat action	Average loss from implementation	Number of means and methods of neutralization	Notes
1	2	3	4	5	6	7	8
1	Password theft or password search.	Internal and external	1. Tracking the authorized user 2. Obtaining a password from a user file or from a system OS file, magnetic media or other source (notebook, smart card, etc.) by unauthorized or legal access or theft. 3. Full search of password variants. 4. Optimized password search (using special lists of the most common passwords)	Compromising the password of a legal user, appearance of a potential possibility of unauthorized access to the information available to the corresponding user	Potentially high 4–5	3, 6, 11,	If a password theft (search) is not detected, the possibility of the repeated unauthorized access and bug setting is retained

(continued)

Table A.2 (continued)

Item no.	Name	Source of threat	Possible ways of implementation	Threat action	Average loss from implementation	Number of means and methods of neutralization	Notes
	Infection with viruses						
2	Introduction of bugs (Trojan programs)	Internal and external	1. Use of unlicensed software. 2. Running unknown programs by an authorized user. 3. Introduction of Trojan programs from global networks (through legal users)	Possibility of remote control of the infected computer system and unauthorized access to the confidential information	3	9, 8, 6	May lead to more severe consequences, e.g., threat No. 1

(continued)

Table A.2 (continued)

Item no.	Name	Source of threat	Possible ways of implementation	Threat action	Average loss from implementation	Number of means and methods of neutralization	Notes
3	Infecting BS elements with viruses	Internal and external	1. Use of unlicensed software, files, and media from unreliable sources. 2. Using infected files from external and global networks	Declining quality of service (deterioration of temporal indicators), loss of control over information flows, violation of confidentiality of information and, in some cases, denial of service	3	9, 8, 6, 4 (as confirmation of reliability of information and sources)	May lead to more severe consequences

Unauthorized access without infringement of authority

(continued)

(continued)

Table A.2 (continued)

Item no.	Name	Source of threat	Possible ways of implementation	Threat action	Average loss from implementation	Number of means and methods of neutralization	Notes
4	Audio interception of information traffic	External and internal	Due to the peculiarities of implementation of some network protocols, the intruder will have access to all information exchange between computers of this segment when connecting to a certain segment of the local network	Breach of confidentiality of information	4–5	1, 2, 5, applying the switching equipment	
5	Collection of gibberish	Internal and external	The intruder reads and analyzes deleted files, swap files, etc.	Compromising passwords, breach of information confidentiality		3, 10	
6	Media scanning	Internal, external	The intruder consistently tries to open files and folders in order to detect network administration errors	Breach of confidentiality of information	3–4	6, 7	

Table A.2 (continued)

Item no.	Name	Source of threat	Possible ways of implementation	Threat action	Average loss from implementation	Number of means and methods of neutralization	Notes
Infringement of authority							
7	Starting the program as a system (driver) or on behalf of the user who has the necessary permissions.	Internal, external	Using errors in the software or OS administration, the intruder obtains powers exceeding those granted to him under the current security policy.	Breach of confidentiality of information, unauthorized modification of information	4	6, 8	
8	Substitution of a dynamically loaded library used by system programs or change of environment variables describing the path to such libraries.	External, internal	By obtaining privileged access to system resources, the intruder introduces a modified version of the system library into the system	Breach of confidentiality of information, unauthorized modification of information	4	6, 8	

(continued)

Table A.2 (continued)

Item no.	Name	Source of threat	Possible ways of implementation	Threat action	Average loss from implementation	Number of means and methods of neutralization	Notes
9	Modifying the code or data of the OS protection subsystem	Internal, external	By obtaining privileged access to OS software and resources, the intruder has the ability to change some of the OS components and protection systems in order to facilitate privileged access to data in the future	Breach of confidentiality of information, unauthorized modification of information	4	6, 8	
Denial of service							
10	Resource locking	Internal, external	The intruder's program locks all the resources available in the OS and then enters an infinite loop	Total or partial OS failure	2–3	6, 7	

(continued)

Table A.2 (continued)

Item no.	Name	Source of threat	Possible ways of implementation	Threat action	Average loss from implementation	Number of means and methods of neutralization	Notes
11	Inquiry attack	Internal, external	The intruder's program constantly sends inquiries to the OS, the response to which requires significant resources of the computer system	Decrease in channel capacity, deterioration of temporary system performance indicators	2–3	5, 6, 7	
12	System malfunction	Internal, external	Sending a message that disables a BS element temporarily or for a long period of time	Disabling the system or its components	2–3	5, 6, 7	
13	Using errors in software or administration	Internal, external	Errors in network administration, network software, etc. give the intruder access to confidential information	Violation of information confidentiality, unauthorized change of information, violation of the system functioning process	4–5	5, 6, 7	

(continued)

Table A.2 (continued)

Item no.	Name	Source of threat	Possible ways of implementation	Threat action	Average loss from implementation	Number of means and methods of neutralization	Notes
14	Replay attack	Internal, external	By sending messages to the network with a false network address, the intruder fraudulently switches to his or her computer already established network connections and as a result obtains the rights of users of these connections	Imposing false information, introducing a virus, a Trojan program, disrupting the system's operation	4–5	5, 6, 7	
15	Getting access to the router table (reading)	Internal, external	Getting information from the router table to provide the ability to send false messages (masquerades)	Breach of confidentiality of information, redirection of information flows	4–5	1 (of control traffic), 5, 6, 7	
16	Changing the router table	Internal, external	Getting privileged access to the router in order to change one or more network addresses of its table to redirect the packet flow.	Breach of confidentiality of information, violation of information exchange	4–5	1 (of control traffic), 5, 6, 7	

(continued)

Table A.2 (continued)

Item no.	Name	Source of threat	Possible ways of implementation	Threat action	Average loss from implementation	Number of means and methods of neutralization	Notes
17	Creating a false router	Internal, external	By obtaining privileged access to system resources, the intruder creates a false router and is capable of intercepting all messages passing through the router, and while it is impossible to capture all the messages that pass through the router due to the volume of information that is too large, selective interception of messages containing users' passwords and e-mail is a huge attraction	Breach of confidentiality of information, violation of information exchange	4–5	1 (of control traffic), 5, 6, 7	

Table A.3 List of information security tools and methods

Item No.	Name	Cost	Efficiency	Notes
1	Packet encryption (traffic)	2500	High	Path, VIPNet, Continent, Thorn, Needle
2	Means of encrypting individual messages (mail, transmitted files, blocks)	250	High	Secure VIPNet mail service, Courier
3	Cryptographic protection of information on magnetic media		High	Cryptomania, StrongDisk
4	Tools for electronic digital signature		High	CryptoPro CSP, VCERT PKI
5	Firewalls	From 1,500 to 40,000.	Average	Zastava-Jet, FortE+ (Zastava-Elvis), FW-1, Cisco PIX, Black Hole (SecurlT Fire-Wall), Cyber Guard, AltaVista
6	Audit/recording tools (security analysis)	7505 (segment) 720 (for 1 comp.) 1535 (for 1 segment)	Average	RealSecure Network Sensor RealSecure OS Sensor RealSecure Manager fr Open View
7	Security scanners	1439 (for 10 dev.) 1001 (for 1 dev.) 2,868 (1-100 users per server)	Above average	Internet Scanner System Scanner Server Database Scanner for Oracle
8	Integrity controls	1000	High	
9	Antivirus tools	30 and above (for one user)	Low	Dialogue Science DrWeb, AVP Kaspersky Labs, NAV for Windows 9x v5.0 Symantec, VirusScan v4.0.0 McAfee[1]
10	Tools for guaranteed file deletion		High	Shredder programs
11	Countermeasures against password search			As a rule, the T&P software settings are used

Table A.4 Results of solution of the task of optimization of the protection system

No. of option	Total cost of the protection system	Total losses caused by the intruder's actions	Probability of accomplishment of all goals by the intruder	Probability of accomplishment of individual goals			
				Goal No. 1	Goal No. 2	Goal No. 3	Goal No. 4
1	1190	8156	0.106	0.4063	0.706	0.548	0.679
2	2627	5220	0.032	0.407	0.7	0.16	0.678
3	4100–5254	5145	0.012	0.217	0.7	0.123	0.651
4	6730–7881	5145	0.00911	0.217	0.7	0.123	0.487
5	10508	5081	0.006067	0.22	0.574	0.071	0.678
6	12200–13135	3993	0.000317	0.029	0.569	0.029	0.651
7	14800–21000	3993	0.000236	0.029	0.569	0.029	0.487
8	23640–25000	3983	0.00012	0.023	0.569	0.014	0.651
9	26300	3983	0.00009	0.0023	0.569	0.014	0.487

Table A.5 Results of solution of the task of optimization of the protection system taking into account the possibility of joint use of protection means

No. of option	Total cost of the protection system	Total losses caused by the intruder's actions	Probability of accomplishment of all goals by the intruder	Probability of accomplishment of individual goals			
				Goal no. 1	Goal no. 2	Goal no. 3	Goal no. 4
1	1190	8156	0.106	0.4063	0.706	0.548	0.679
2	2627	1022	$2.916 \; 10^{-5}$	0.2154	0.0341	0.0691	0.0575
3	4100–5254	366	$5.801 \; 10^{-7}$	0.0251	0.0309	0.0273	0.0276
4	6730–7881	344	$4.802 \; 10^{-7}$	0.0251	0.0309	0.0273	0.0227
5	10508	289	$3.391 \; 10^{-7}$	0.0422	0.0074	0.0064	0.0345
6	12200–13135	199	$4.921 \; 10^{-8}$	0.0175	0.0043	0.0235	0.0247
7	14800–21000	188	$4.016 \; 10^{-8}$	0.0175	0.0096	0.0235	0.0112
8	23640–25000	83	$1.207 \; 10^{-8}$	0.0111	0.0032	0.0082	0.0038
9	26300	83	$1.116 \; 10^{-8}$	0.0111	0.0032	0.0082	0.0038

Table A.6 Composition of the protection system for different options of solution of the corresponding optimization task

No. of threat to information.	No. of means and methods of protection included in the system								
	Option no. 1	Option no. 2	Option no. 3	Option no. 4	Option no. 5	Option no. 6	Option no. 7	Option no. 8	Option no. 9
1	11	11	3.11	3.11	11	3.11	3.11	3.11	3.11
2	9	9	8.9	8.9	9	8.9	8.9	8.9	8.9
3	4	4	4	4	4	4	4	4	4
4	2	2	2	2	–	2	2	2	2
5	10	–	10	10	10	10	10	10	10
6	–	7	7	7	7	7	7	6.7	6.7
7	–	–	(8)	(8)	(6)	(6,8)	(8)	(6,8)	(6,8)
8	–	–	(8)	(8)	(6)	(6,8)	(8)	(6,8)	(6,8)
9	–	–	(8)	(8)	(6)	(6,8)	(8)	(6,8)	(6,8)
10	–	(7)	(7)	(1,7)	(6,7)	(6,7)	(7)	(6,7)	(6,7)
11	–	(7)	(7)	(1,7)	(6,7)	(6,7)	(5,7)	(5,6,7)	(5,6,7)
12	–	(7)	(7)	(1,7)	(6,7)	(6,7)	(5,7)	(5,6,7)	(5,6,7)
13	–	(7)	(7)	(1,7)	6 (7)	6 (7)	5 (7)	5 (6,7)	5 (6.7)
14	–	(7)	(7)	(1,7)	(6,7)	(6,7)	(5,7)	5 (6,7)	5 (6,7)
15	–	(7)	(7)	1 (7)	(6,7)	(6,7)	1 (5,7)	(5,6,7)	1 (5,6,7)
16	–	(7)	(7)	(1,7)	(6,7)	(6,7)	(1,5,7)	(5,6,7)	(1,5,6,7)
17	-	(7)	(7)	(1,7)	(6,7)	(6,7)	(1,5,7)	(5,6,7)	(1,5,6,7)

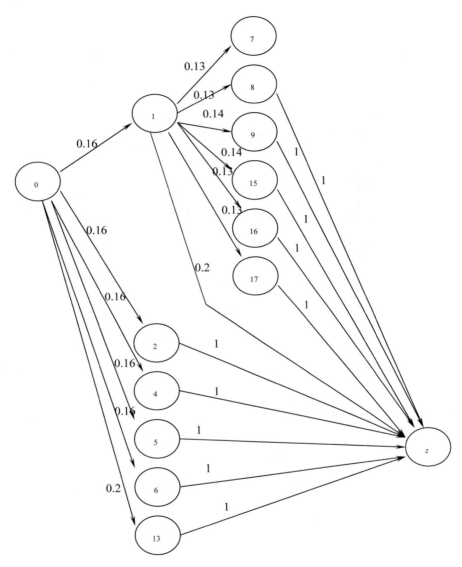

Fig. A.1 State bars for the description of the intruder's target No 1. The vertex number 0 corresponds to the initial state of the system when none of the threats is implemented. The state with the number z corresponds to the implementation of the target by the intruder. The remaining vertices of the bars correspond to the threats being implemented

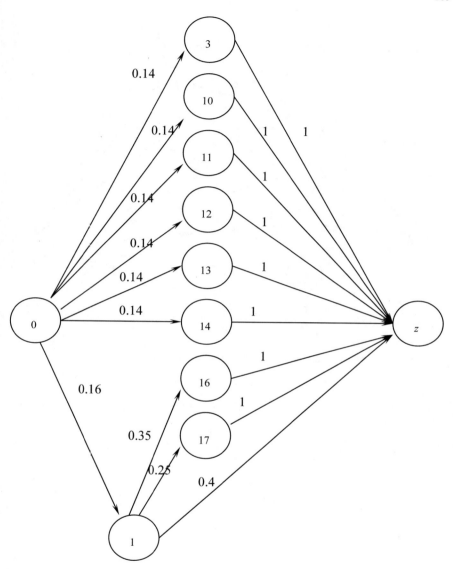

Fig. A.2 System state bars for the description of the intruder's target no. 2

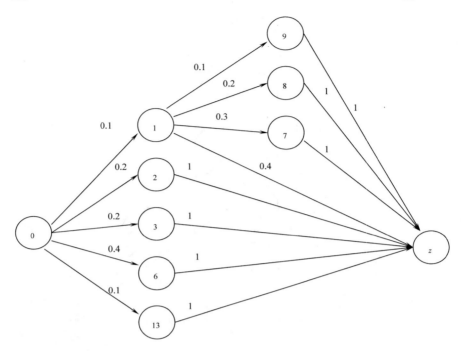

Fig. A.3 System state bars for the description of the intruder's target no. 3

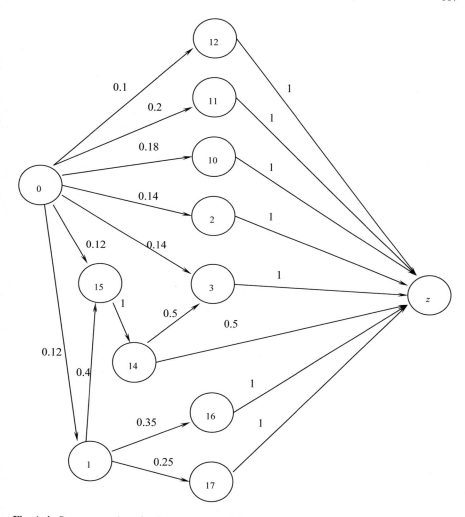

Fig. A.4 System state bars for the description of the intruder's target no. 4

Appendix B

Data Channel Selection Program
  Certificate of state registration of the computer program
  No. 2018664835
  Date of request receipt October 01, 2018
  Date of state registration in the Register of computer programs November 22, 2018.

D. A. Zatuchny et al., *Noise Resistance Enhancement in Aircraft Navigation and Connected Systems*, Springer Aerospace Technology,
https://doi.org/10.1007/978-981-16-0630-4

Language: JavaScript

```javascript
$(function() {
    var funcs = [];
    var systems = [];
    function updateSystemCount() {
        var sysCount = $("#sysCount");
        sysCount.html(systems.length);
    }
    updateSystemCount();
    var $funcTab = $("#initFunt");
    var $systTab = $("#initSyst");
    var $choSys = $("#chooseSyst");
    function unique(arr) {
        var obj = {};
        for (var i = 0; i < arr.length; i++) {
            var str = arr[i];
            obj[str] = true;
        }
        return Object.keys(obj);
    }
    $("#addNewFunk").on('click', function () {
        var $wrap = $(".functions-block");
        var template = '<label class="form-label func-template"> Enter the name
of the function that the system should perform (Example: Working with
passengers) ' +
                        '<input type="text" class="form-control system-func">'
+
                        '</label>';
```

```javascript
    $wrap.append(template);
  });
  $("#createFuncs").on('click', function () {
    var $funcBlock = $(".system-func");
    funcs = [];
    var error = false;
    $funcBlock.each(function () {
      if($(this).val().length < 1){
        alert("Fill in the required fields! ");
        error = true;
      }
    });
    if(!error){
      $funcBlock.each(function () {
        var value = $(this).val();
        funcs.push(value);
      });
      funcs = unique(funcs);
      $funcTab.hide("medium");
      $systTab.show("medium");
    }
  });
  $("#addNewSystemBtn").on('click', function () {
    $(".system__array").html("");
    $(".system__inner").each(function () {
      $(this).find('input[type="checkbox"]').eq(0).prop('checked', true);
      $(this).find('input[type="radio"]').eq(0).prop('checked', true)
    });
    var $systems = $(".systems");
    $systems.html("");
```

```javascript
//var data = $("#template").clone();
$("#template").clone().appendTo($systems);
//$systems.append(data);
console.log(funcs);
for(var i = 0; i < funcs.length; i++){
    var chClass = "";
    if(i === 0) {
        chClass = "checked";
    }
    var template = ' <label class="function-check" data-fid="'+i+'">' +
funcs[i] + ':' +
        '<input type="checkbox" class="input-function" ' + chClass + '
title="'+funcs[i]+'" data-fid="'+i+'">' +
        '</label>';
    $(".system__array").append(template);
}
});
$("body").on('keyup', '[name=level]', function () {
    var $c = $(this);
    if($c.val() < 0) {
        $c.val(0);
    }else if($c.val() > 100){
        $c.val(100);
    }else {
        $c.val($c.val());
    }
});
```

```
$('body').on('click', '.saveSystemBTN', function () {
  var $systems = $(".system__array");
  var checkedFuncs = [];
  var chekCount = 0;
  $systems.find('.input-function').each(function () {
    console.log($(this));
    console.log($(this).prop('checked'));
    console.log($(this).val());
    console.log(chekCount);
    console.log("----");
    if($(this).prop('checked') === true) {
      chekCount++;
      var currFunc = {
        id: $(this).attr('data-fid'),
        val: $(this).attr('title')
      };
      checkedFuncs.push(currFunc);
    }
  });
  console.log("Saved features ");
  //unique(checkedFuncs);
  checkedFuncs.splice(0, 1);
  console.log(checkedFuncs);
  console.log("---");
  var errors = false;
  var title = $(".input-title").val();
  if(title.length < 1){
    alert("Specify the name of the system!");
    errors = true;
```

```
        }
    if(chekCount < 1) {
        alert("Select the functionality of the system! ");
        errors = true;
    }
    if(!errors){
        var system = {
            title: title,
            funcs: checkedFuncs,
            endStand: $('[name="standart"]:checked').val(),
            orvd: $('[name="garmon"]:checked').val(),
            level: $('[name="level"]]').val(),
            cost: $('[name="cost"]]').val(),
            rusAdapt: $('[name="have-russ"]:checked').val(),
            evo: $('[name="have-evo"]:checked').val()
        };
        systems.push(system);
        alert("The system was successfully added! ");
        $(".systems").html("");
        updateSystemCount();
    }
});
$("#createSyst").on('click', function(){
    if(systems.length >= 2){
        $systTab.hide("medium");
        //
        var $fuList = $(".systems__allowed-func-list");
        $fuList.html("");
        for(var i = 0; i < funcs.length; i++){
```

```
      var li = "<li class='systems__analyse-fulitem'><label class='form-
label'>"+funcs[i]+" :<input type='checkbox' value='"+funcs[i]+"' data-
fid='"+i+"' class='in-sy'/></label></li>";
        $fuList.append(li);
      }
      $choSys.show("medium");
    }else {
      alert("Insufficient systems for analysis! Please add at least 2 systems!
");
    }
  });
  $("#podbor").on('click', function(){
    var $li = $(".systems__analyse-fulitem");
    var checked = 0;
    var needleFun = [];
    $li.each(function () {
      var $in = $(this).find('input');
      if($in.prop('checked') === true) {
        checked++;
        var fun = {
          id: $in.attr('data-fid'),
          val: $in.val()
        };
        needleFun.push(fun);
      }
    });
    if(checked === 0){
      alert("Select at least one function that the system must perform! ");
      return false;
```

```
}else {
  console.log(needleFun);
  console.log(systems);
  var $analyseBlock = $(".systems__analyse");
  $analyseBlock.html("");
  for(var i = 0; i < systems.length; i++){
    var isFuncs = false;
    var isStd = false;
    var isORVD = false;
    var isLevel = false;
    var isCost = false;
    var isNoRf = false;
    var isNoPrem = false;
    var curFuncs = systems[i]['funcs'];
    for(var j = 0; j < curFuncs.length; j++){
      for(var k = 0; k < needleFun.length; k++){
        if(curFuncs[j]['id'] === needleFun[k]['id']){
          isFuncs = true;
        }
      }
    }//--
    if(curFuncs.length < needleFun.length) {
      isFuncs = false;
    }
    if(systems[i]['endStand'] === "Да"){
      isStd = true;
    }
    if(systems[i]['orvd'] === "Да"){

    isORVD = true;
```

```
   }
   console.log("System Level: " + parseInt(systems[i]['level']));
   console.log("Required level: "+ parseInt($("#needLVL").val()));


   if( parseInt(systems[i]['level']) >= parseInt($("#needLVL").val())){
      isLevel = true;
   }


   if( parseInt(systems[i]['cost']) <= parseInt($("#needCOST").val())){
      isCost = true;
   }
   if(systems[i]['rusAdapt'] === "Yes"){
      isNoRf = true;
   }
   if(systems[i]['evo'] === "Yes"){
      isNoPrem = true;
   }
   var funStr = "";


   for(var kek = 0; kek < systems[i]['funcs'].length; kek++){
      funStr += "<li>" + systems[i]['funcs'][kek]['val']+ "</li>";
   }
   var wrnClass = "";
   var errorDesc = "";
   var descStr = "";
   if(isFuncs === false){
      wrnClass = "is-none";
      errorDesc += " - No functionality required <br/>";
   }
```

```
if(isStd === false){
    wrnClass = "is-hz";
    var stCost = 1000;
    errorDesc += "<br/> - The process of international standardization
is not over. Options:<br/> -- The system is not used<br/> -- The process of
international standardization is coming to an end soon and the system is being
used (the cost is increased by" + stCost + " rub)<br/><br/>";
    var kekes = parseInt(systems[i]['cost']);
    kekes += stCost;
    systems[i]['cost'] = kekes;
}
if(isORVD === false){
    wrnClass = "is-hz";
    var stCost = 5000;
    errorDesc += "<br/> - There is no factor in harmonizing the
services of ATM systems in Russia, the United States and Europe. Variants:
<br/> - The system is not used <br/> - The system is used on the basis of
temporary arrangements <br/> The system is being finalized for the identity
of the services being presented, which leads to an increase in the cost of its
operation by "+ stCost +" rubles<br/><br/>";
    var kekes = parseInt(systems[i]['cost']);
    kekes += stCost;
    systems[i]['cost'] = kekes;
}
if(isLevel === false){
    wrnClass = "is-hz";
    errorDesc += "<br/> - Level of system performance does not
correspond to the required level. Variants: <br/> The system is not used
<br/> - The system is being finalized in the near future with the increase in
cost and is used. <br/> <br/>";
}
```

```
if(isCost === false){
    wrnClass = "is-hz";
    errorDesc += "<br/> - The cost of using the system exceeds the
allowable limit. Variants: <br/> The system is not used <br/> The system is
used in especially necessary cases. <br/> <br/>;
}
if(isNoRf === false){
    wrnClass = "is-hz";
    errorDesc += "<br/> - The system is not adaptable to the
conditions of the Russian Federation.Variants: <br/> - The system is not used
<br/> - The system is used in areas with increased traffic intensity as an
additional independent source of surveillance. <br/> <br/>;
}
if(isNoPrem === false){
    wrnCl   ass = "is-hz";
    errorDesc += "<br/> - The system does not satisfy the condition of
continuity. Options: <br/> - The system is not used <br/> the system can be
used in the future when it will satisfy the condition of continuity. <br/> - The
system is used as a backup option. <br /> <br />;
}
if(!isFuncs){
    wrnClass = "is-none";
    //errorDesc = "- No functionality required<br/>";
}
if(isFuncs && isStd && isORVD && isLevel && isCost && isNoRf
&& isNoPrem){
    wrnClass = "is-ok";
}
var tmp = '<div class="col-sm-6">'+
            '<div class="systems__analyseblock '+wrnClass+'">' +
            '<div class="systems__analyse
```

```
title">'+systems[i]['title']+'</div>' +
                    '<div class="systems__analyseerror">' + errorDesc
+'</div>' +
                    '<div class="systems__analysefuncs">' +
                    '<span class="systems__analyse
span">Functional:</span>'+
                        '<ul class="systems__analysefunlist">' + funStr +
'</ul>' +
                    '</div>' +
                    '<div class="systems__analysestand"> Completeness
of the international standardization process: <span>' systems[i]['endStand']
+'</span></div>'+
                    '<div class="systems__analysestand">
Harmonization of ATM services in Russia, the United States and Europe:
<span>'+systems[i]['orvd']+'</span></div>'+
                    '<div class="systems__analysestand"> Level of
technology (%): <span>'+ parseInt(systems[i]['level'])+'%</span></div>'+
                    '<div class="systems__analysestand"> Cost  (rub):
<span>'+ parseInt(systems[i]['cost'])+'</span></div>' +
                    '<div class="systems__analysestand"> Adaptability
to Russia: <span>'+ systems[i]['rusAdapt'] +'</span></div>'+
                    '<div class="systems__analysestand"> Evolutionary
and continuity of the process of creating and implementing AZPB  in Russia:
<span>'+ systems[i]['evo'] +'</span></div>'+
                    '</div>' +
                    '</div>';
            $analyseBlock.append(tmp);
        }
        $(".systems__analyseblock ").matchHeight();
    }
});
});
```

Printed in the United States
by Baker & Taylor Publisher Services